Control Systems Engineering

Control Systems Engineering

Xavier Taylor

NY RESEARCH PRESS
New York

Published by NY Research Press
118-35 Queens Blvd., Suite 400,
Forest Hills, NY 11375, USA
www.nyresearchpress.com

Control Systems Engineering
Xavier Taylor

International Standard Book Number: 978-1-64725-430-8 (Hardback)

Cataloging-in-Publication Data

Control systems engineering / Xavier Taylor.
p. cm.
Includes bibliographical references and index.
ISBN 978-1-64725-430-8
1. Automatic control. 2. Control theory. 3. Systems engineering. I. Taylor, Xavier.
TJ213 .M63 2023
629.8--dc23

Contents

Preface

Control systems engineering refers to a field of engineering that deals with the principles of control theory in order to create systems that produce the desired behaviors in a controlled way. It focuses on the design and analysis of systems in order to increase the stability, response speed, and accuracy of the system. Control systems engineering necessitates an extensive skill set that encompasses mechanical, electrical and software systems. The most widely used control systems include systems for controling temperature in a building, the speed of a conveyor belt in a process plant, and chemical concentrations in drinking water. This book outlines the processes and applications of control systems engineering in detail. It will also provide interesting topics for research, which interested readers can take up. This book is an essential guide for both academicians and those who wish to pursue this discipline further.

This book is a comprehensive compilation of works of different researchers from varied parts of the world. It includes valuable experiences of the researchers with the sole objective of providing the readers (learners) with a proper knowledge of the concerned field. This book will be beneficial in evoking inspiration and enhancing the knowledge of the interested readers.

In the end, I would like to extend my heartiest thanks to the authors who worked with great determination on their chapters. I also appreciate the publisher's support in the course of the book. I would also like to deeply acknowledge my family who stood by me as a source of inspiration during the project.

Xavier Taylor

1

Mathematical Modeling of Control Systems

1.1 Classification of Control Systems

System

An interconnection of elements and devices for a desired purpose is termed as system.

Control System

Control system is the interconnection of components forming a system configuration that will provide a desired response.

Process

Process is the device, plant or system under control. The input and output relationship represents the cause and effect relationship of the process.

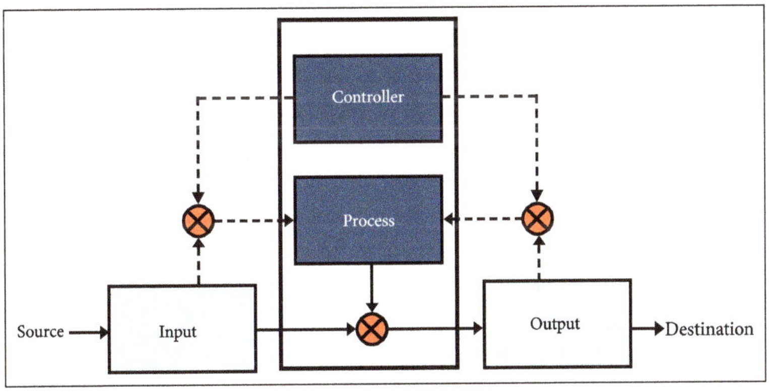

Control system.

Control systems are ubiquitous. They appear in cars, homes, industry and in systems for communication and transport. Control is increasingly becoming as a critical mission, processes will fail if the control does not work.

Control is inherently multidisciplinary. The control system contains actuators, sensors, computers and software. Analysis of the design of control systems require s knowledge about the particular process to be controlled, knowledge of the techniques of control and specific technology used in the sensors and actuators.

Controllers are typically implemented using digital computers. Also knowledge about real time computing and software is therefore essential. Sensors and actuators are often connected by communication networks.

There are two main branches of control systems:

- Open-loop systems.
- Closed-loop systems.

1.1.1 Open Loop and Closed Loop Control Systems and their Differences

Open-loop Systems

The open-loop system is also termed as the non-feedback system. In this open-loop system, there is no way to ensure that the actual l speed is close to the desired speed. The actual speed might be way off the desired speed because of the wind speed or road conditions, such as uphill or downhill etc.

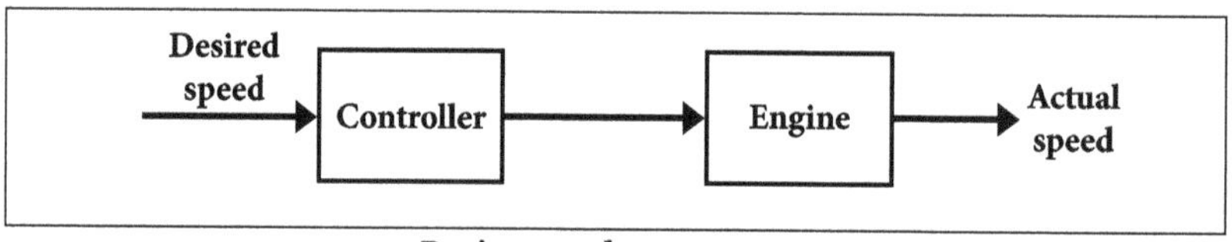

Basic open-loop system.

Closed-loop Systems

The closed-loop system is also called as feedback system. It has the mechanism to ensure that the actual speed is close to the desired speed.

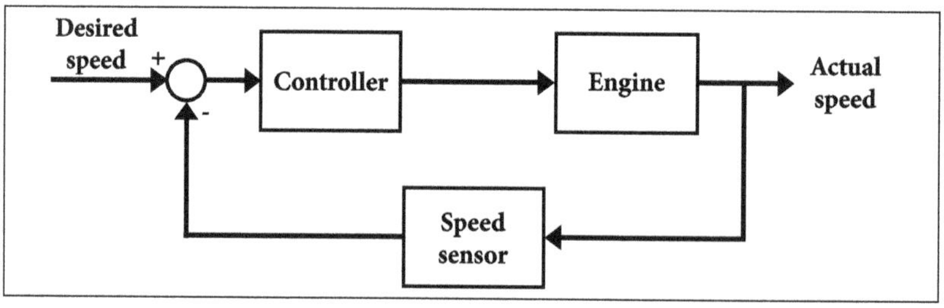

Basic closed-loop system.

Advantages of the closed loop control system:

- Accurate and reliable.
- System is less affected by noise.

- The changes i n output due to the external disturbances are corrected automatically.
- Bandwidth range is large.

Difference between Open Loop and Closed Loop System

Open Loop System	Closed Loop System
Open Loo p systems are generally stable.	Efforts are needed to design a stable system.
Simple and economical.	Complex and costlier.
Inaccurate.	Accurate.
Unreliable.	Reliable.
The changes in the output due to the external disturbances are not corrected automatically.	The changes in the output due to the external disturbances are corrected automatically.

1.2 Feedback Characteristics

In general, feedback is the process by which a fraction of the output signal, either a voltage or a current is used as an input. Feedback in an electronic system might be a negative feedback or a positive feedback but it is unilateral in direction. Its signal flow is one way i.e., only from the output to the input of the system. This makes the loop gain, G of the system independent of the load and source impedances.

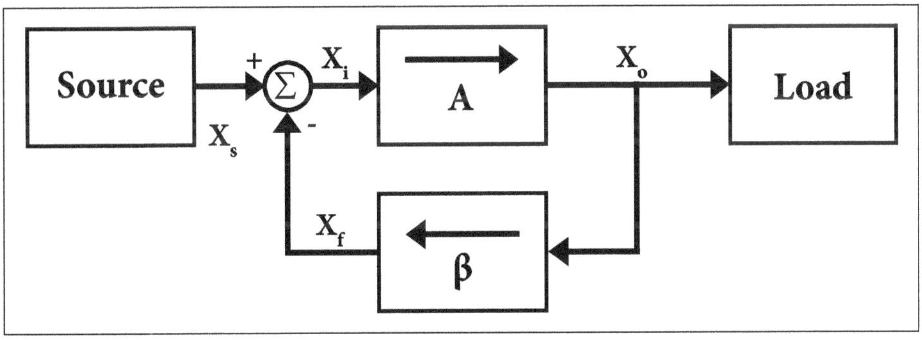

General feedback structure.

If the feedback signal is out of phase with the input signal, then the input voltage applied to the basic amplifier is decreased and correspondingly the output is decreased. This type of feedback is known as negative or degenerative feedback.

The below circuit represents a system with positive gain G and feedback β. The summing junction at its input subtracts the feedback signal from the input signal to form the error signal V_{in} - βG, which drives the system.

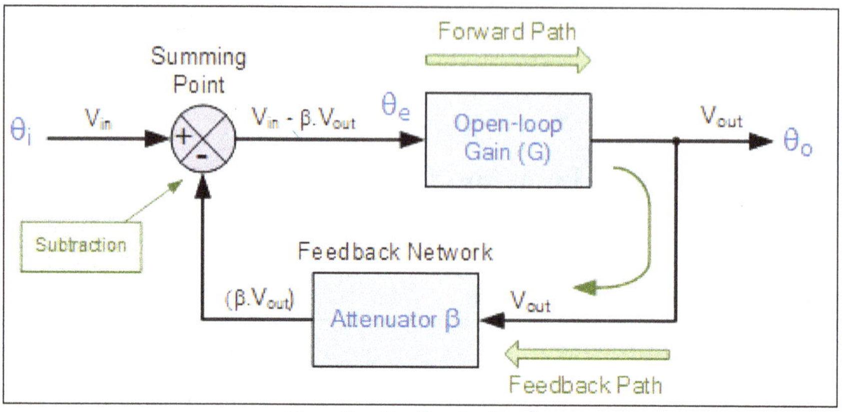

Negative feedback circuit.

We can derive the general feedback equation as,

$$\text{System Gain, } G = \frac{V_{out}}{V_{in}}$$

$$G \times V_{in} = V_{out}$$

$$G\left(V_{in-\ \beta Vout}\right) = V_{out}$$

$$G.V_{in} - G.\beta.V_{out} = V_{out}$$

$$G.V_{in=Vout}\left(1+G\beta\right)$$

$$\therefore \quad \frac{V_{out}}{V_{in}} = Gv = \frac{G}{1+G\beta}$$

This is the negative feedback equation.

Where,

$G =$ Open loop voltage gain

$\beta =$ Feedback fraction

$G\beta =$ Loop gain

$1+G\beta =$ Feedback factor

$Gv =$ Closed loop voltage gain

Negative feedback occurs when some function s of the output of a system, process or mechanism is fed back in a manner which tends to reduce the fluctuations in the output, whether caused by the changes in the input or by other disturbances.

Properties of Negative Feedback

- It increases the stability of the amplifier.
- It reduces the gain of an amplifier.
- It decreases noise and distortion.
- It increases the bandwidth.

Effects of Feedback System

- Gain.
- Stability.
- Sensitivity.
- Noise.

Let the system have open loop gain [G(S)], feedback loop gain [H(S)], output signal [C(S)] and input signal [R(S)]. Then the feedback signal [B(S)] is represented as,

$$B(S)=H(s).C(S)$$

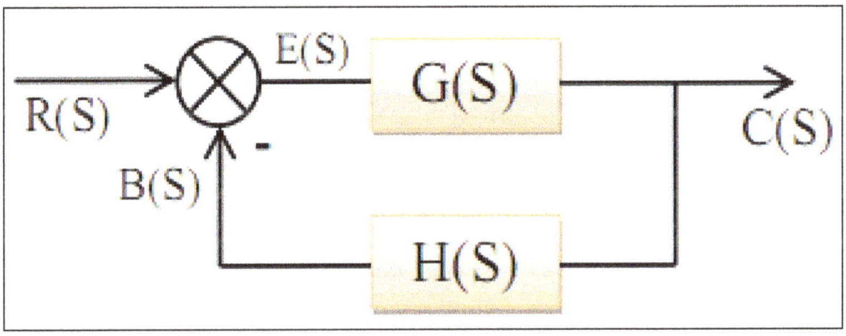

Feedback systems.

$$G(S)=\frac{C(S)}{E(S)} \& \; E(S)=R(S)-B(S)$$

Hence,

$$C(S)=G(S)\,.\,E(S)$$

$$=G(S)\left[R(S)-B(S)\right]$$

$$=G(S)\left[R(S)-H(S).C(S)\right]$$

$$\frac{C(S)}{R(S)}=\frac{G(S)}{1+G(S).H(S)} \qquad ...(1)$$

Gain: Gain of the open loop system is reduced by a factor $\left[1+G(s).\,H(s)\right]$ in the feedback system. Here, the feedback signal is negative. If the feedback gain has positive value, the overall gain will be reduced. If the feedback gain has negative value, the overall gain will be increase d.

Stability: If a system follow s the input command signs then the system is said to be s table. A system is said to be unstable, if its output is out of control.

If G (H) = - 1, the output of the system is infinite for any finite input. This shows that the stable system may become unstable for certain values of the feedback gain. Thus, if the feedback is not properly used, the system can be harmful.

Sensitivity: Sensitivity depends on the system parameters. For a good control system, it is desirable that the system should be insensitive to its parameter changes.

$$\text{Sensitivity, } S_G=\frac{1}{1+GH}$$

Sensitivity of the system can be reduced by increasing the value of G (H). This can be done by selecting proper feedback.

Noise: The effect of feedback on the noise signals will be greatly influenced by the point at which the signals are introduced in the system. It is possible to reduce the effect of noise by proper designing of the feedback system.

Examples: Brush and commutation noise in electrical machines.

1.3 Transfer Function of Linear System

Transfer function is one type of modeling a system. Using first principle, differential equation is obtained. Laplace transform can be applied to the equation assuming zero initial conditions. Ratio of LT (output) to LT (input) is expressed as a ratio of polynomial in s in the transfer function.

Transfer function of a LTIV system is defined as the ratio of the Laplace Transform of the output variable to the Laplace Transform of the input variable assuming all the initial condition as zero.

Properties of Transfer Function

- The transfer function of a system is the mathematical model expressing the differential equation that relates the output to input of the system.
- The transfer function is the property of a system independent of magnitude and the nature of the input.
- The transfer function includes the transfer functions of the individual elements. But at the same time, it does not provide any information regarding physical structure of the system.
- The transfer functions of many physically different systems shall be identical.
- If the transfer function of the system is known, the output response can be studied for various types of inputs to understand the nature of the system.
- If the transfer function is unknown, it may be found out experimentally by applying known inputs to the device and studying the output of the system.

Methods to Obtain the Transfer Function (T.F.)

- Write the differential equation of the system.
- Take the Laplace Transform of the differential equation, assuming all the initial condition to be zero.
- Take the ratio of the output to the input. This ratio is the T.F.

Problems

1. Let us obtain the transfer function for the below network.

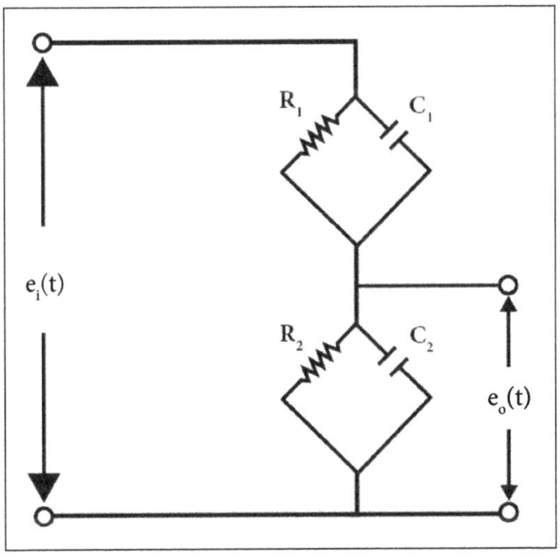

Solution:

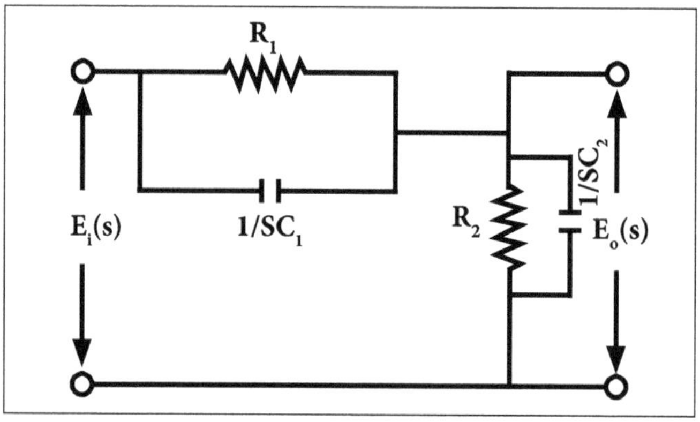

$$E_o(s) = \frac{R_2 \| 1/SC_2}{(R_1 \| 1/SC_4) + (R_2 \| 1/SC_2)} \times E_i(s)$$

$$\frac{E_o(s)}{E_i(s)} = \frac{\left(\dfrac{R_2/SC_2}{R_2 + \dfrac{1}{SC_2}}\right)}{\dfrac{R_1}{R_1 + \dfrac{1}{SC_1}} + \dfrac{R_2/SC_2}{\left(R_2 + \dfrac{1}{SC_2}\right)}}$$

$$= \frac{\dfrac{R_2}{1+SC_2R_2}}{\dfrac{R_1}{1+SC_1R_1} + \dfrac{R_2}{1+SC_2R_2}}$$

$$= \frac{\dfrac{R_2}{(1+SC_2R_2)}}{\dfrac{R_1(1+SC_2R_2)+R_2(1+SC_1R_1)}{(1+SC_1R_1)(1+SC_2R_2)}}$$

$$= \frac{R_2(1+SC_1R_1)}{R_1(1+SC_2R_2)+R_2(1+SC_1R_1)}$$

$$\frac{E_o(s)}{E_i(s)} = \frac{R_1(1+SC_1R_1)}{R_1+R_2+2SC_2R_2R_1}$$

The above circuit can be drawn in Laplace domain as,

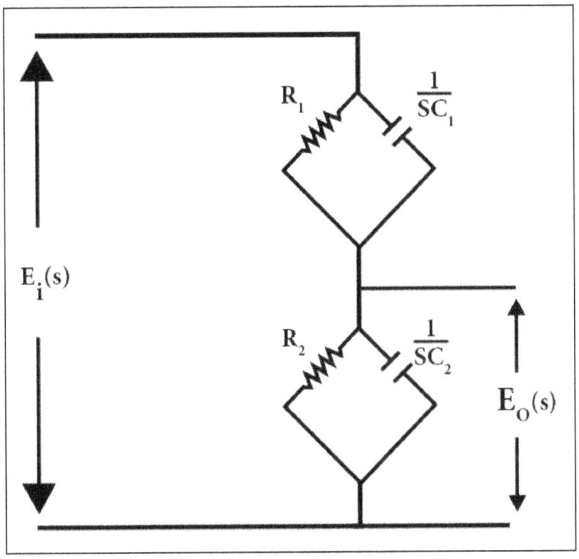

Then by voltage division rule,

$$E_o(s)=\frac{\dfrac{R_2/C_2s}{R_2+\dfrac{1}{C_2s}}}{\dfrac{R_2/C_2s}{R_2+\dfrac{1}{C_2s}}+\dfrac{R_1/C_1s}{R_1+\dfrac{1}{C_1s}}}*E_i(s)$$

$$\frac{E\ (s)}{E\ (s)}\quad\frac{\overline{R_2C_2s\ \ 1}}{\dfrac{R_2}{R_2C_2s+1}\quad\dfrac{R_1}{R_1C_1s+1}}$$

$$\frac{E_o(s)}{E_i(s)}=\frac{\dfrac{R_2}{R_2C_2s+1}}{\dfrac{R_2(R_1C_1s+1)+R_1(R_2C_2s+1)}{(R_2C_2s+1)(R_1C_1s+1)}}$$

$$\frac{E_o(s)}{E_i(s)}=\frac{R_2(R_1C_1s+1)}{R_2(R_1C_1s+1)+R_1(R_2C_2s+1)}$$

2 . Let us derive the transfer function for the below diagram.

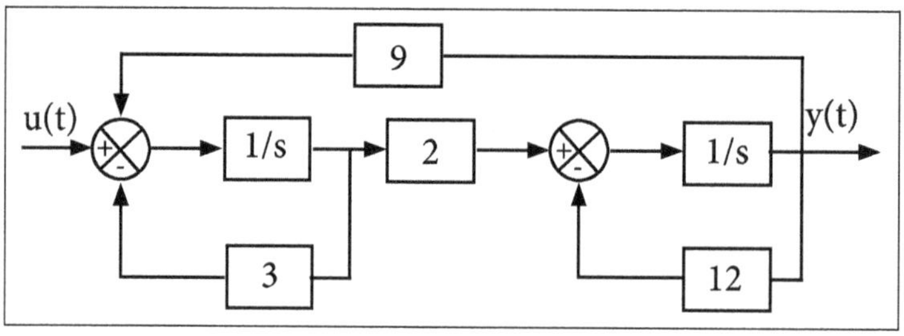

Solution:

Step 1: Reduce the feedback loops.

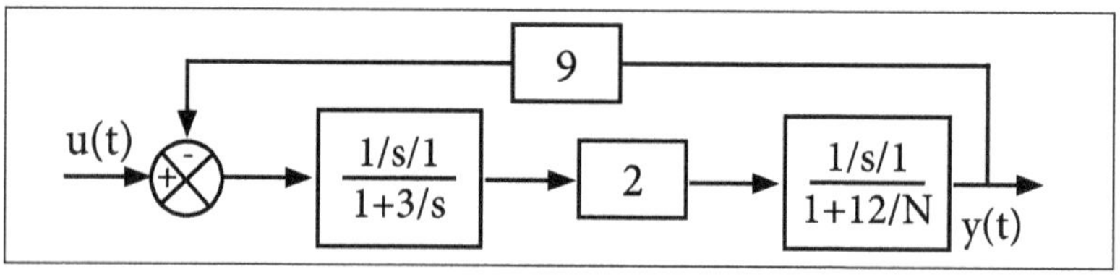

Step 2: Cascade the blocks.

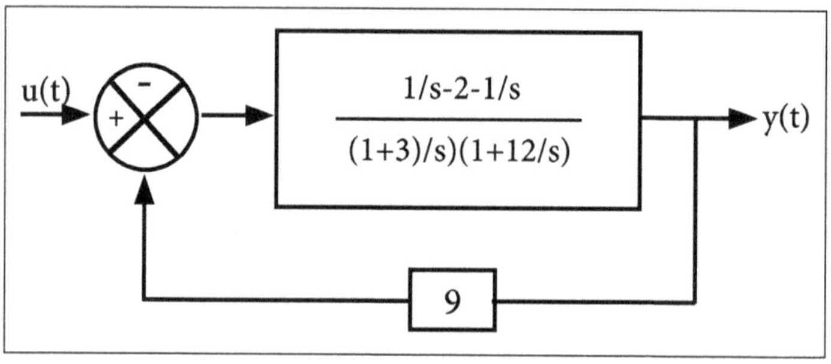

$$G = \frac{1/s \cdot 2 \cdot 1/s}{(1+3/s)(1+12/s)} = \frac{2/s^2}{\left(\frac{s+3}{s}\right)\left(\frac{s+12}{s}\right)}$$

$$\Rightarrow \frac{\frac{2}{s^2}}{\frac{(s+2)(s+12)}{s^2}} \Rightarrow \frac{2}{(s+3)(s+12)}$$

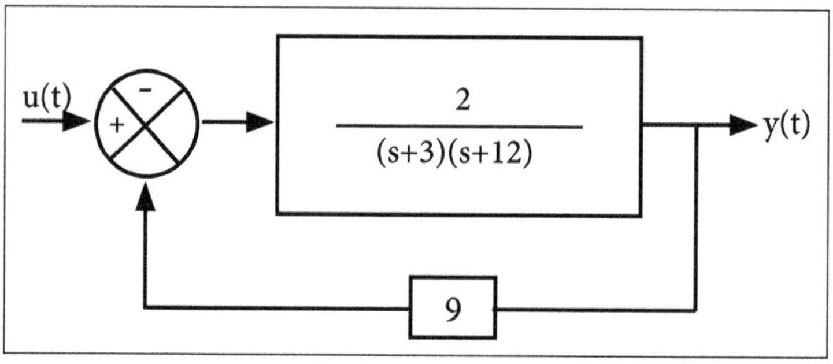

Step 3: Reduce the feedback loop.

$$\frac{y(t)}{U(t)}=\frac{\frac{2}{(s+3)(s+12)}}{1+\frac{2\times 9}{(s+3)(s+12)}}=\frac{2}{\frac{(s+3)(s+12)+18}{(s+3)(s+12)}}$$

$$\frac{y(t)}{U(t)}=\frac{2}{(s+3)(s+12)+18}$$

3. Let us derive the transfer function for the system whose block diagram is as shown below.

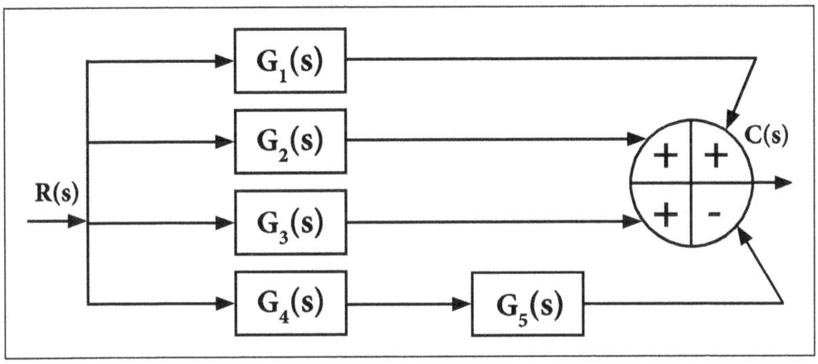

Solution:

To find: Transfer function,

$$\frac{C(s)}{R(s)}=G_1(s)+G_2(s)+G_3(s)-G_4(s)G_5(s).$$

4. Let us determine the transfer function for the below network.

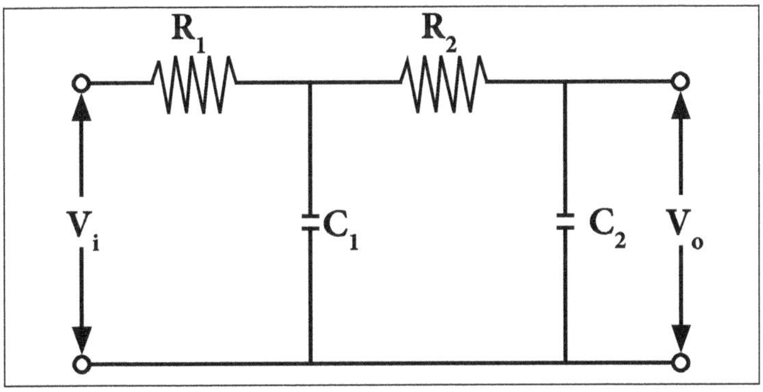

Solution:

To find: Transfer function,

Formula to be used:

$$T \cdot F = \frac{V_o(s)}{V_i(s)}$$

Where,

V_o (s) = Voltage drop around C_2.

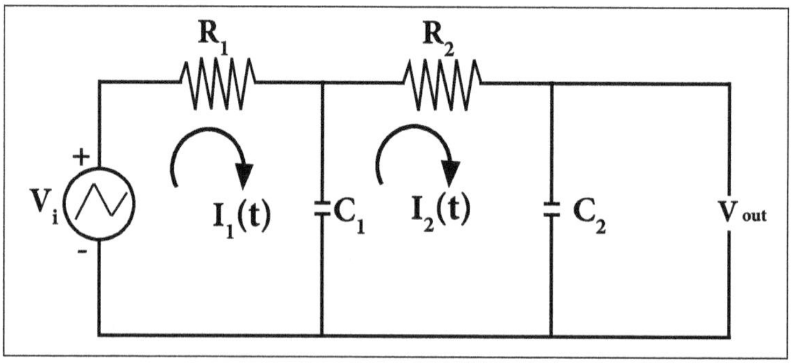

$$V_i(s) = R_1 I_1(s) + \frac{1}{sC_1}\left[I_1(s) - I_2(s)\right]$$

$$V_i(s) = \left[R_1 + \frac{1}{sC_1}\right] I_1(s) - \frac{1}{sC_1} I_2(s)$$

Determine KVL, for loop 2.

$$R_2 I_2(s) + \frac{1}{sC_2} I_2(s) + \frac{1}{sC_1}\left[I_2(s) - I_1(s)\right] = 0$$

$$\Rightarrow -\frac{1}{sC_1} I_1(s) + \left[R_2 + \frac{1}{sC_1} + \frac{1}{sC_2}\right] I_2(s) = 0.$$

$$\begin{bmatrix} V_i(s) \\ 0 \end{bmatrix} = \begin{bmatrix} R_1 + \frac{1}{sC_1} & -1/sC_1 \\ -1/sC_1 & R_2 + \frac{1}{sC_1} + \frac{1}{sC_2} \end{bmatrix} \begin{bmatrix} I_1(s) \\ I_2(s) \end{bmatrix}$$

$$I_2(s) = \frac{\Delta_2}{\Delta}$$

$$\Delta = \left[R_1 + \frac{1}{sC_1}\right]\left[R_2 + \frac{1}{sC_1} + \frac{1}{sC_2}\right] - \frac{1}{s^2 C_1^2}$$

$$=R_1R_2+R_1\left[\frac{1}{sC_1}+\frac{1}{sC_2}\right]+\frac{R_2}{sC_1}+\frac{1}{s^2C_1C_2}$$

$$\Delta_2=\begin{vmatrix} R_1+\frac{1}{sC_1} & V_i \\ -1/sC_1 & 0 \end{vmatrix};\ \Delta^2=\frac{V_i(s)}{sC_1}$$

$$I_2(s)=\frac{\frac{V_i(s)}{sC_1}}{\frac{R_1R_2C_1C_2S^2+(R_1C_1+R_1C_2+R_2C_2)S+1}{s^2C_1C_2}}$$

$$V_o(s)=\frac{I_2(s)}{sC_2}$$

$$\therefore\ T\cdot F=\frac{V_o(s)}{V_i(s)}$$

$$=\frac{1}{R_1R_2C_1C_2S^2+(R_1C_1+R_1C_2+R_2C_2)S+1}.$$

5. Let us determine the output C(s) due to R(s) and disturbance D(s) for the below figure.

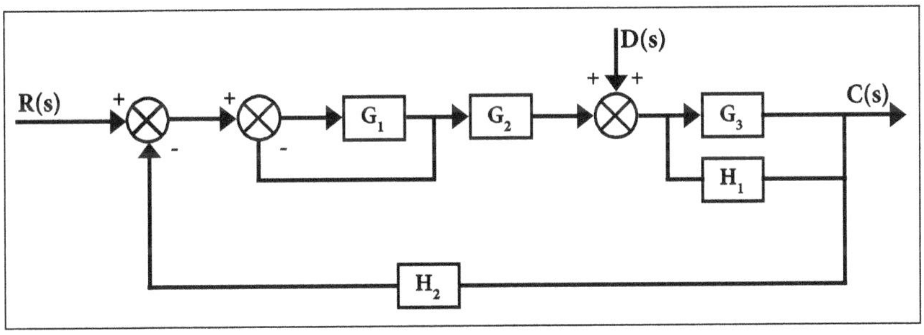

Solution:

To find: Transfer function,

$$\frac{C(s)}{R(s)}=\frac{G_1G_2G_3}{(1+G_1)(1+G_3H_1)+G_1G_2G_3H_2}.$$

$$\frac{C(s)}{R(s)}=\frac{(1+G_1)G_3}{(1+G_1)(1+G_3H_1)+G_1G_2G_3H_2}$$

6. Let us obtain the transfer function of the following electrical network.

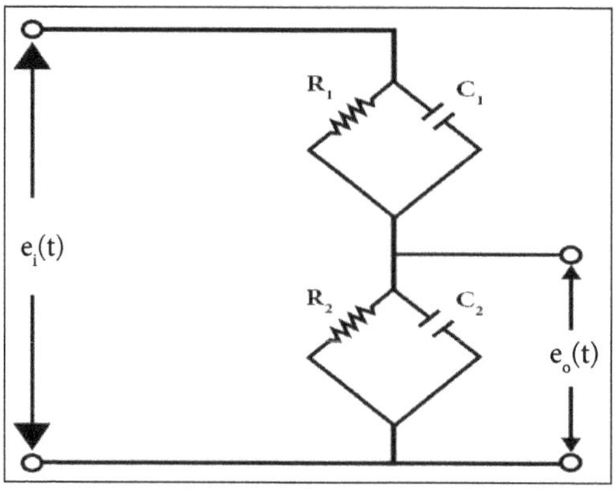

Solution:

Given:

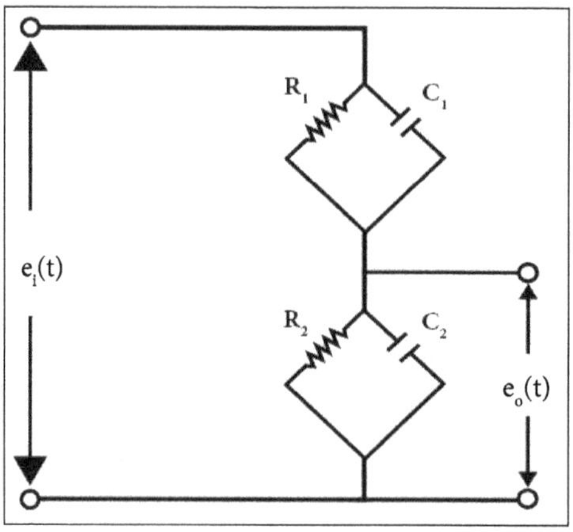

To find: Transfer function,

The above circuit can be drawn in Laplace domain, then by voltage division rule,

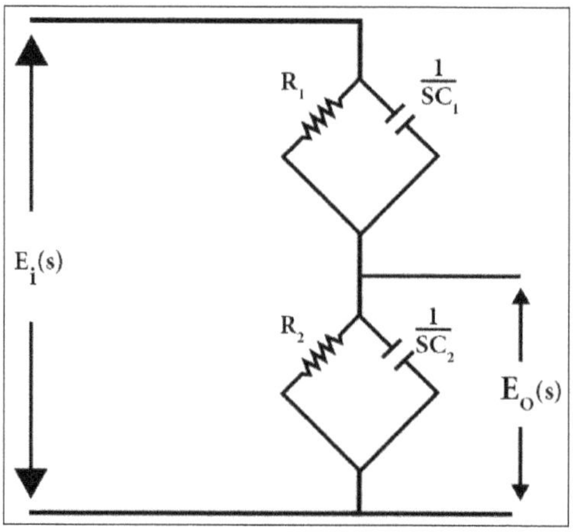

The Transfer function is given by,

$$\frac{e_1(s)}{e_o(s)}=\frac{R_2(R_1C_1s+1)}{R_2(R_1C_1s+1)+R_1(R_2C_2s+1)}$$

Note: Convert the diagram into Laplace transform, and then use voltage divider formula to find $e_{1(s)}$.

Converting circuit into Laplace is nothing but use same circuit, replace e(t) to e(s) and C to $1/C_s$, L to L_s.

$$E_o(s)=\frac{\dfrac{R_2/C_2s}{R_2+\dfrac{1}{C_2s}}}{\dfrac{R_2/C_2s}{R_2+\dfrac{1}{C_2s}}+\dfrac{R_1/C_1s}{R_1+\dfrac{1}{C_1s}}}*E_i(s)$$

$$\frac{E_o(s)}{E_i(s)}=\frac{\dfrac{R_2}{R_2C_2s+1}}{\dfrac{R_2}{R_2C_2s+1}+\dfrac{R_1}{R_1C_1s+1}}$$

$$\frac{E_o(s)}{E_i(s)}=\frac{\dfrac{R_2}{R_2C_2s+1}}{\dfrac{R_2(R_1C_1s+1)+R_1(R_2C_2s+1)}{(R_2C_2s+1)(R_1C_1s+1)}}$$

$$\frac{E_o(s)}{E_i(s)}=\frac{R_2(R_1C_1s+1)}{R_2(R_1C_1s+1)+R_1(R_2C_2s+1)}$$

7. Let us find the transfer function for the below network.

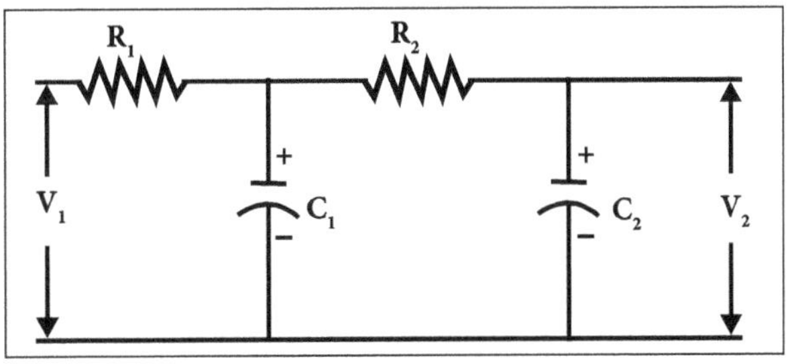

Solution:

To find: Transfer function,

In Laplace domain, the circuit can be drawn as,

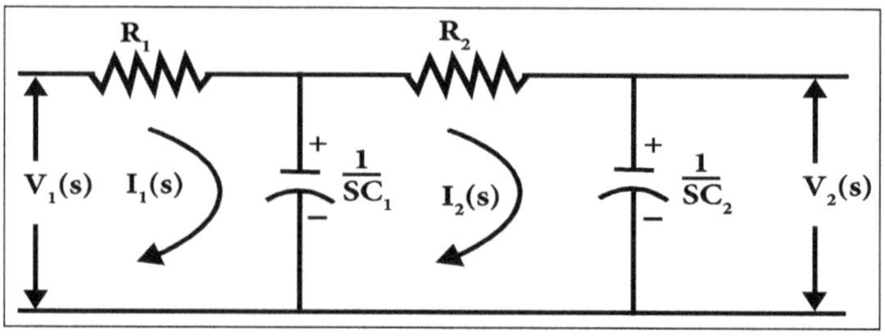

There are two loops in the above circuit by writing Mesh equations,

$$\begin{bmatrix} V_i(s) \\ 0 \end{bmatrix} = \begin{bmatrix} R_1 + \frac{1}{sC_1} & -\frac{1}{sC_1} \\ -\frac{1}{sC_1} & R_2 + \frac{1}{sC_2} + \frac{1}{sC_2} \end{bmatrix} \begin{bmatrix} I_1(s) \\ I_2(s) \end{bmatrix}$$

$$\begin{bmatrix} V_i(s) \\ 0 \end{bmatrix} = \left[R + \frac{1}{C_1 s} \right] I_1(s) - \frac{1}{C_1 s} I_2(s)$$

$$0 = -\frac{1}{C_1 s} I_1(s) + \left[R + \frac{1}{C_2 s} + \frac{1}{C_1 s} \right] I_2(s) \Rightarrow I_1(s) = C_1 s = \left[R + \frac{1}{C_2 s} + \frac{1}{C_1 s} \right] I_2(s)$$

Substitute this in the above equation,

$$V_i(s) = \left[R_1 + \frac{1}{C_1 s} \right] C_1 s \left[R_2 + \frac{1}{C_2 s} + \frac{1}{C_1 s} \right] I_2(s) - \frac{1}{C_1 s} I_2(s)$$

$$= \left\{ \left[\frac{R_1 C_1 s + 1}{C_1 s} \right] \left[\frac{R_2 C_1 C_2 s^2 + C_1 s + C_2 s}{C_1 C_2 s^2} \right] C_1 s - \frac{1}{C_1 s} \right\} I_2(s)$$

$$= \left[\frac{(R_1 C_1 s + 1)(R_2 C_1 C_2 s + C_1 + C_2)}{C_1 C_2 s} - \frac{1}{C_1 s} \right] I_2(s)$$

$$= \left[\frac{(R_1 C_1 s + 1)(R_2 C_1 C_2 s + C_1 + C_2)}{C_1 C_2 s} \right] I_2(s)$$

The output voltage is given by,

$$V_2(s) = \frac{1}{C_2 s} I_2(s) = \left[\frac{1}{C_2 s} \times \frac{C_1 C_2 s}{(R_1 C_1 s + 1)(R_2 C_1 C_2 s + C_1 + C_2) - C_2} \right] V_i(s)$$

8. Let us calculate the transfer function $Y_2(s)/F(s)$ for the below system.

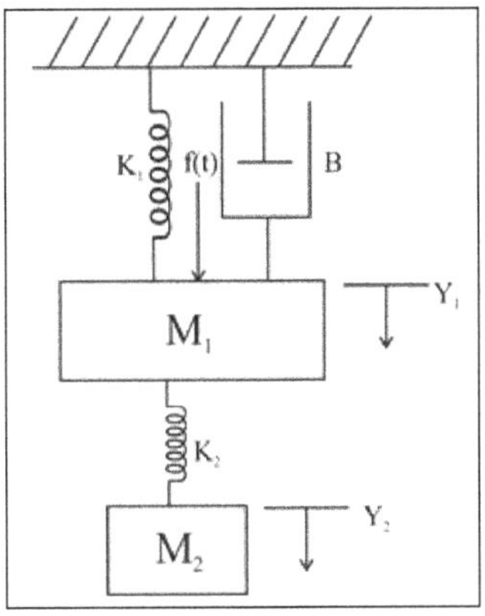

Solution:

To find:

Transfer function: $Y_{2(s)/F(s)}$

Force acting on M_1 is given by,

$$F(t)=M_1\frac{d^2y_1}{dt}+B\frac{dy_1}{dt}+K_1 y_1(t)+K_2\left[y_1(t)-y_2(t)\right] \quad ...(1)$$

Force acting on M_2 is given by,

$$0=M_2\frac{d^2y_2}{dt}+K_2\left[y_2(t)-y_1(t)\right] \quad ...(2)$$

Take Laplace Transform of above equation, then,

$$F(s)=[M_1s^2+Bs+(K_1+K_2)]Y_1(s)-K_2Y_2(s) \quad ...(3)$$

$$0=[M_2s^2+K_2]Y_2(s)-K_2Y_1(s) \quad ...(4)$$

From equation (4),

$$[M_2s^2+K_2]Y_2(s)=K_1Y_1(s)$$

$$Y_1(s)=\frac{\left[M_2s^2+K_2\right]Y_2(s)}{K_2}$$ Substitute this in equation (3).

$$F(s)=\left[M_1s^2+Bs+(K_1+K_2)\right]\frac{\left[M_2s^2+K_2\right]}{K_2}Y_2(s)-K_2Y_2(s)$$

$$F(s)=\left(\frac{\left[M_1s^2+Bs+(K_1+K_2)\right]\left[M_2s^2+K_2\right]-K_2^2}{K_2}\right)Y_2(s)$$

$$\frac{Y_2(s)}{F(s)} = \frac{K_2}{\left\{\left[M_1 s^2 + Bs + (K_1 + K_2)\right]\left[M_2 s^2 + K_2\right] - K_2^2\right\}}$$

1.4 Differential Equations of Electrical Networks

Differential Equations

- A simpler system or element may be governed by first order or second order differential equation. When several elements are connected in sequence, say "n" elements, each one with first order, the total order of the system will be nth order.

- In general, a collection of components or system shall be represented by nth order differential equation.

$$a_0 y(t) + a_1 \frac{d}{dt} y(t) + \ldots + a_{n-1} \frac{d^{n-1}}{dt_{n-1}} y(t) + a_n \frac{d^n}{dt^n} y(t) - b_0 x(t) - b_1 \frac{d}{dt} x(t) + \ldots$$

$$+ b_{n-1} \frac{d^{n-1}}{dt^{n-1}} x(t) + \ldots + b_{n-1} \frac{d^{n-1}}{dt^{n-1}} x(t) + b_n \frac{d^n}{dt^n} x(t)$$

- In control systems, transfer function characterizes the input output relationship of components or systems that can be described by Liner Time Invariant Differential Equation.

- In the earlier period, the input output relationship of a device was represented graphically.

- In a system having two or more components in sequence, it is very difficult to find graphical relation between the input of the first element and the output of the last element. This problem is solved by transfer function.

Electrical Systems

Electrical circuits involving resistors, capacitors and inductors are considered. The behavior of such systems are governed by Ohm's law and Kirchhoff's laws.

Resistor

Consider a resistance of 'R' Ω carrying current 'i' Amps as shown in figure, then the voltage drop across it is V = RI.

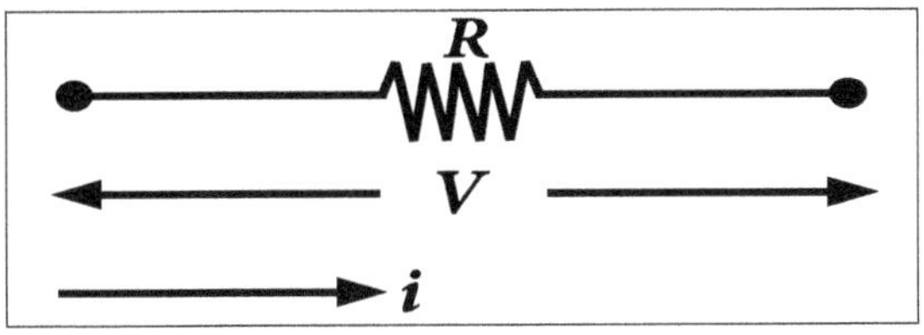

Inductor

Consider an inductor 'L' H carrying current 'i' Amps as shown in figure, then the voltage drop across it can be written as V = L di / dt.

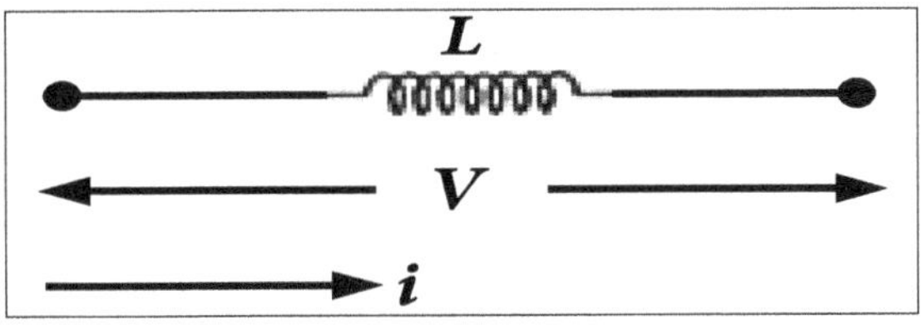

Capacitor

Consider a capacitor 'C' F carrying current 'i' Amps as shown in figure, then the voltage drop across it can be written as $V = (1/C)\int i\,dt$.

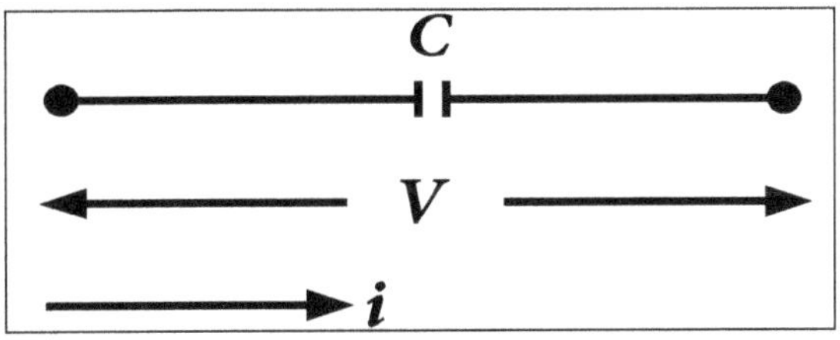

Steps for modeling of electrical system:

- Apply Kirchhoff's voltage law or Kirchhoff's current law to form the differential equations describing electrical circuits comprising of resistors, capacitors and inductors.
- Form Transfer Functions from the describing differential equations.
- Then simulate the model.

Example:

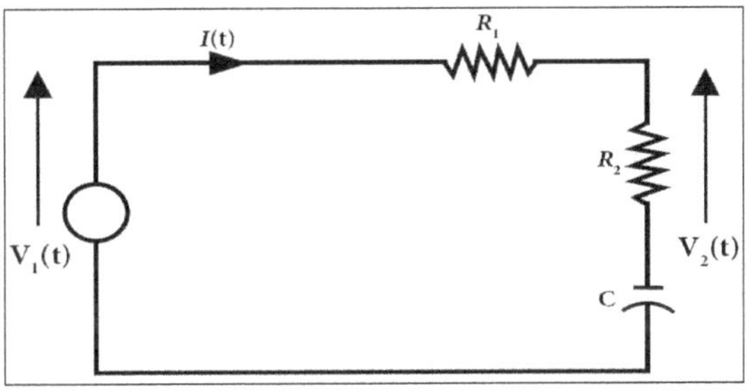

$$R_1 i(t) + R_2 i(t) + 1/C \int i(t)\, dt = V(t) R_2 i(t) + 1/C \int i(t)\, dt = V(t)$$

Analogous Systems

Let us consider a mechanical (both translational and rotational) and electrical system as shown in the figure:

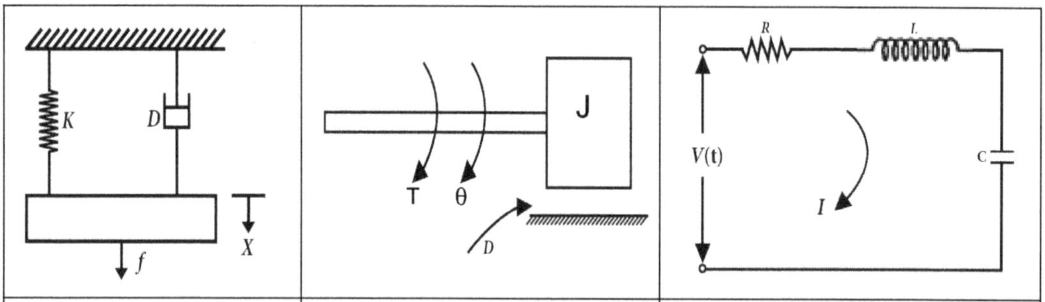

(a) Translational (b) Rotational (c) Electrical system

From figure (a), we get

$$M\, d^2x/dt^2 + D\, dx/dt + K_x = f$$

From figure (b), we get

$$M\, d^2\theta/dt^2 + D\, d\theta/dt + K\theta = T$$

From figure (c), we get

$$L\, d^2q/dt^2 + R\, dq/dt + (1/C)q = V(t)$$

Where $q = \int i\, dt$

They are two methods to get analogous system. These are:

- Force-voltage (f-v) analogy.
- Force-current (f-c) analogy

Force–voltage (f-v) Analogy

Translational	Electrical	Rotational
Force(f)	Voltage(v)	Torque(T)
Mass(M)	Inductance(L)	Inertia(J)
Damper(D)	Resistance(R)	Damper(D)
Spring(K)	Elastance (1/C)	Spring(K)
Displacement(x)	Charge(q)	Displacement(θ)
Velocity(u)	Current(i)	Velocity(ω)

Force–current (f-c) Analogy

Translational	Electrical	Rotational
Force(f)	Current(i)	Torque(T)
Mass(M)	Capacitance (C)	Inertia(J)
Spring(K)	Reciprocal of inductance(1/L)	Damper(D)
Damper(D)	Conductance(1/K)	Spring(K)
Displacement(x)	Flux Linkage(ψ)	Displacement(θ)
Velocity $(u = dx / dt)$	Voltage $(v) = d\psi / dt$	Velocity $(\omega = d\theta / dt)$

Problems

1. Let us derive analogous electrical system for the below mechanical system using force-voltage analogy.

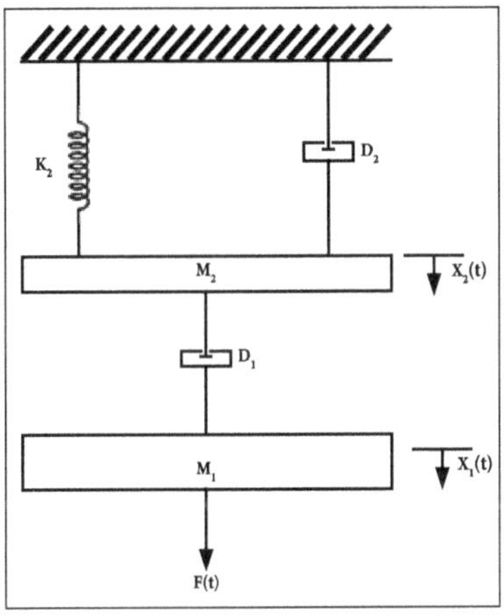

Solution:

To find: Analogous electrical system,

Formula to be used:

Force – Voltage Analogy,

$f(t) = e(t);$

$B = R;$

$M = L$ and $K = 1/C;$

Displacement x (t) = Electric charge q (t)

Force acting on M_1,

$$F(t)=M_1\frac{d^2x_1(t)}{dt^2}+D_1\left[\frac{dx_1(t)}{dt}-\frac{dx_2(t)}{dt}\right]$$

Force acting on M_2,

$$0=M_2\frac{d^2x_2(t)}{dt^2}+D_1\left[\frac{dx_2(t)}{dt}-\frac{dx_1(t)}{dt}\right]+D_2\left[\frac{dx_2(t)}{dt}\right]+k_2x_2(t)$$

By using Force: Voltage Analogy,

The parameters for F-V analogy are f (t) = e (t); B = R; M = L and K = 1/ C; displacement x (t)= electric charge q (t).

Substituting these parameters in the above equation, we get,

$$e(t)=L_1\frac{d^2q_1(t)}{dt^2}+R_1\left[\frac{dq_1(t)}{dt}-\frac{dq_2(t)}{dt}\right]$$

$$0=L_2\frac{d^2q_2(t)}{dt^2}+R_1\left[\frac{dq_2(t)}{dt}-\frac{dq_1(t)}{dt}\right]+R_2\left[\frac{dq_2(t)}{dt}\right]+\frac{1}{C_2}q_2(t)$$

We know that $i(t)=\frac{dq(t)}{dt}$, then above equation becomes

$$e(t)=L_1\frac{di_1(t)}{dt^2}+R_1\left[i_1(t)-i_2(t)\right]$$

$$0=L_2\frac{di_2(t)}{dt}+R_1\left[i_2(t)-i_1(t)\right]+R_2i_2(t)+\frac{1}{C_2}\int i_2(t)dt$$

From the above equation, Force-Voltage analogy can be drawn as,

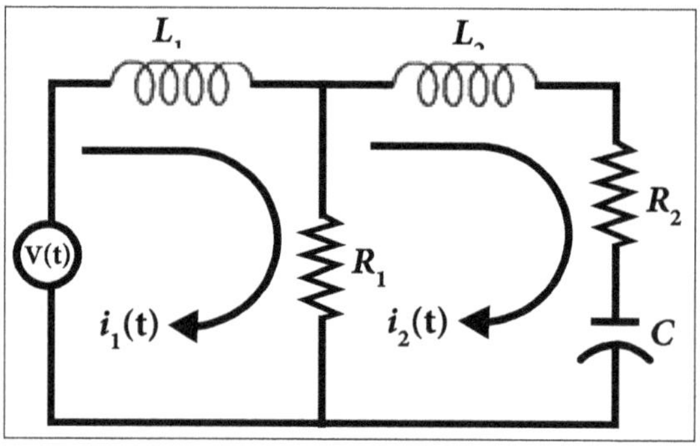

2. Let us draw the Force-Voltage and Force-Current electrical analogous circuits for the below system and verify it through mesh and node equations.

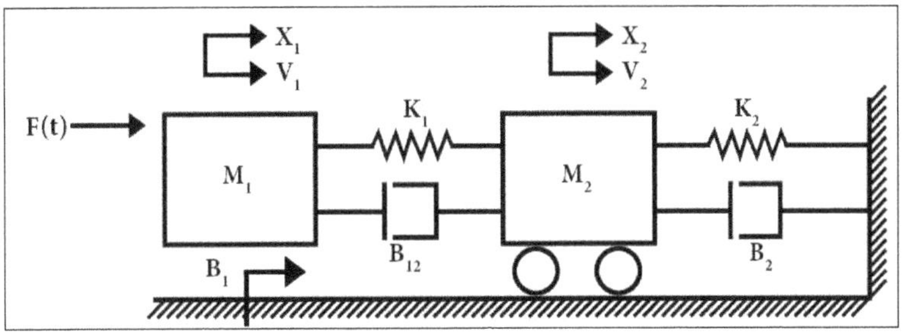

Solution:

To find:

Force voltage and force current electrical analogous circuit.

For mass M_1:

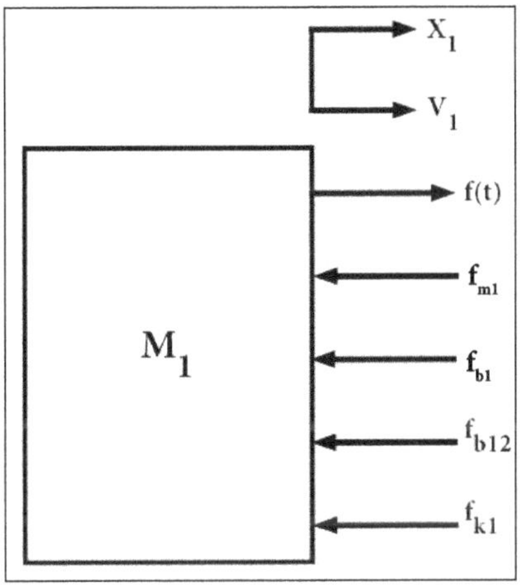

By Newton's second law,

$$f_{m1} + f_{b1} + f_{b12} + f_{k1} = f(t)$$

$$M_1 \frac{d^2x_1}{dt^2} + B_1 \cdot \frac{dx_1}{dt} + B_{12} \frac{d}{dt}(x_1 - x_2) + K_1(x_1 - x_2) = f(t) \qquad ...(1)$$

For mass M_2:

By Newton's second law,

$$f_{m2} + f_{b2} + f_{k2} + f_{b12} + f_{k1} = 0$$

$$M_2 \cdot \frac{d^2x_2}{dt^2} + B_2 \cdot \frac{dx_2}{dt} K_2x_2 + B_{12} \frac{d}{dt}(x_2 - x_1) + K_1(x_2 - x_1) = 0 \qquad ...(2)$$

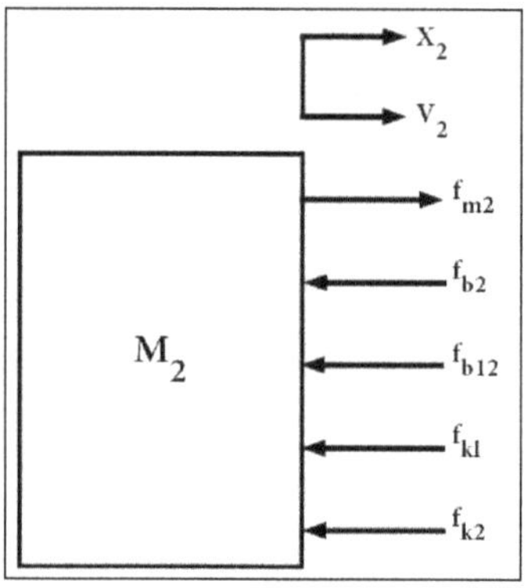

On replacing the displacement of velocity in the differential equations, we have,

$$\text{i.e.,}\frac{d^2x}{dt^2}=\frac{dv}{dt};\frac{dx}{dt}=V;x=\int V\,dt$$

$$M_1\cdot\frac{dv_1}{dt}+B_1v_1+B_{12}(v_1-v_2)K_1\int(v_1-v_2)dt=f(t) \quad ...(3)$$

$$M_2\cdot\frac{dv_2}{dt}+B_2v_2+K_2\int v_2dt+B_{12}(v_2-v_1)+K_1\int(v_2-v_1)dt=0 \;...(4) \quad ...(4)$$

Force-Voltage Analogous Circuit:

The electrical analogous elements for the elements of mechanical system are as follows,

$f(t)\rightarrow e(t)$	$M_1\rightarrow L_1$	$B_1\rightarrow R_1$	$K_1\rightarrow\frac{1}{C_1}$
$V_1\rightarrow i_1$	$M_2\rightarrow L_2$	$B_2\rightarrow R_2$	$K_2\rightarrow\frac{1}{C_2}$
$V_2\rightarrow i_2$		$B_{12}\rightarrow R_{12}$	

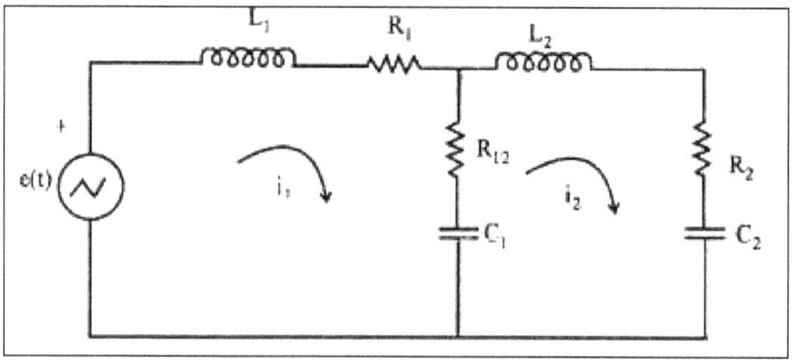

Force-Voltage analogous circuit.

The mesh basic equations using KVL for the circuit are,

$$L_1\frac{di_1}{dt}+R_1i_1+R_{12}(i_1-i_2)+\frac{1}{C_1}\int(i_1-i_2)dt=e(t)$$

$$L_2\frac{di_2}{dt}+R_2 i_2+\frac{1}{C_2}\int i_2\times dt+R_{12}(i_2-i_1)+\frac{1}{C_1}\int(i_2-i_1)dt=0.$$

Force-Current Analogous Circuit:

The electrical analogous elements for the elements of mechanical system are as follows:

$f(t)\rightarrow i(t)$	$M_1\rightarrow C_1$	$B_1\rightarrow 1/R_1$	$K_1\rightarrow 1/L_1$
$V_1\rightarrow V_1$	$M_2\rightarrow C_2$	$B_2\rightarrow 1/R_2$	$K_2\rightarrow 1/L_2$
$V_2\rightarrow V_2$		$B_{12}\rightarrow 1/R_{12}$	

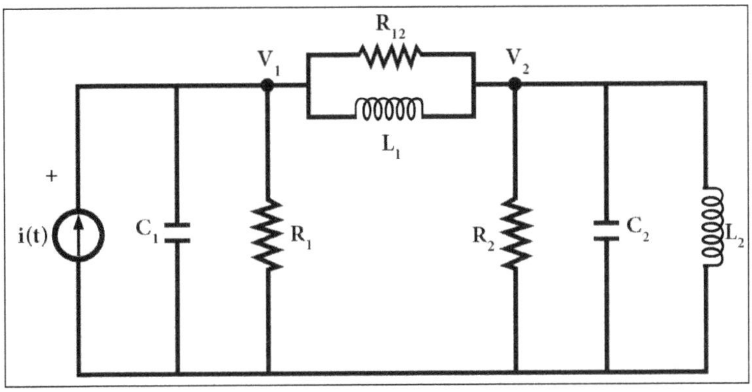

Force-Current analogous circuit.

The node basis equations using KCL for the circuit is,

$$C_1\cdot\frac{dV_1}{dt}+\frac{1}{R_1}V_1+\frac{1}{R_{12}}(V_1-V_2)+\frac{1}{L_1}\int(V_1-V_2)^{dt=i(t)}$$

$$C_2\cdot\frac{dV_2}{dt}+\frac{1}{R_2}V_2+\frac{1}{L_2}\int V_2\,dt+\frac{1}{R_{12}}(V_2-V_1)+\frac{1}{L_1}\int(V_2-V_1)dt=0$$

1.5 Translational and Rotational Mechanical Systems

Mechanical Translational Systems

The model of mechanical translational systems can obtain by using three basic elements such as mass, spring and dashpot. When a force is applied to a translational mechanical system, it is opposed by opposing forces due to mass, friction and elasticity of the system.

The force acting on a mechanical body is governed by Newton's second law of motion. For translational systems, it states that the sum of forces acting on a body is zero.

Force Balance Equations of Idealized Elements

1. Consider an ideal mass element shown in the below figure, which has negligible friction and elasticity. Let a force be applied on it. The mass will offer an opposing force which is proportional to acceleration of a body.

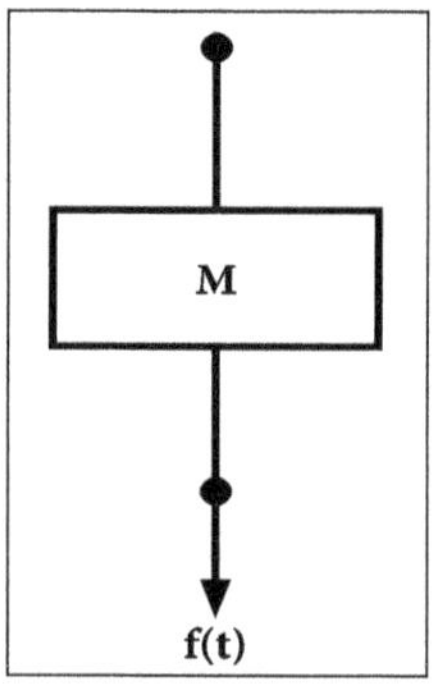

Mass element.

Let,

f = Applied force.

f_m = Opposing force due to mass.

Here,

$f_m \alpha M d^2x / dt^2$

By Newton's second law,

$f = f_m = M d^2x / dt^2$

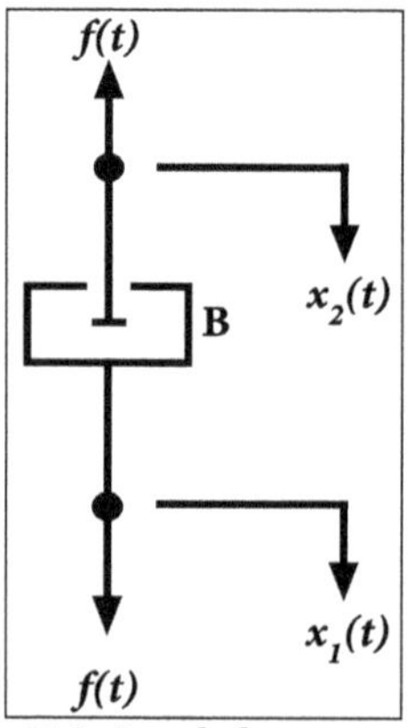

Frictional element.

2. Consider an ideal frictional element dash-pot shown in the above figure has negligible mass and elasticity. Let a force be applied on it. The dashpot will offer an opposing force which is proportional to the velocity of the body.

Let,

f = Applied force.

Fb = Opposing force due to friction.

Here,

$f_b \alpha B dx/dt$

By Newton's second law, $f = f_b = M dx/dt$

3. Consider an ideal elastic element springshown in the below figure. This has negligible mass and friction.

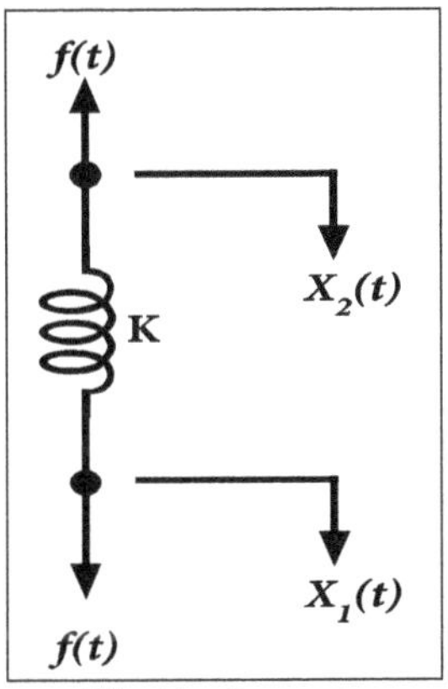

Elastic element.

Let,

f = Applied force.

f_k = Opposing force due to elasticity.

Here,

$f_k \alpha x$

By Newton's second law, $f = f_{k=x}$.

Mechanical Rotational Systems

The model of rotational mechanical systems can be obtained by using three elements such as moment of inertia [J] of mass, dash pot with rotational frictional coefficient [B] and torsional spring with stiffness[k].

When a torque is applied to a rotational mechanical system, it is opposed by opposing torques due to moment of inertia, friction and elasticity of the system. The torque acting on the rotational mechanical bodies is governed by Newton‘s second law of motion for rotational systems.

Torque Balance Equations of Idealized Elements

1. Consider an ideal mass element shown in the below figure has negligible friction and elasticity. The opposing torque due to moment of inertia is proportional to the angular acceleration.

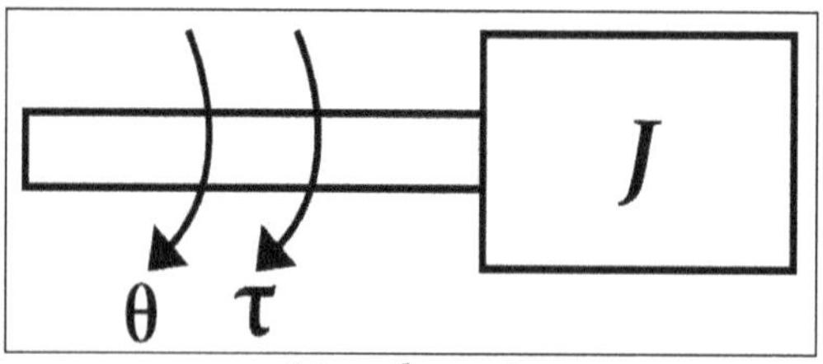

Mass element.

Let,

τ = Applied torque.

τ_j = Opposing torque due to moment of inertia of the body.

Here,

$$\tau_j = \alpha\, J\, d^2\theta / dt^2$$

By Newton‘s law,

$$\tau = \tau_j = J\, d^2\theta / dt^2$$

2. Consider an ideal frictional element dash pot shown in the below figure which has negligible moment of inertia and elasticity. Let a torque be applied on it. The dash pot will offer an opposing torque proportional to the angular velocity of the body.

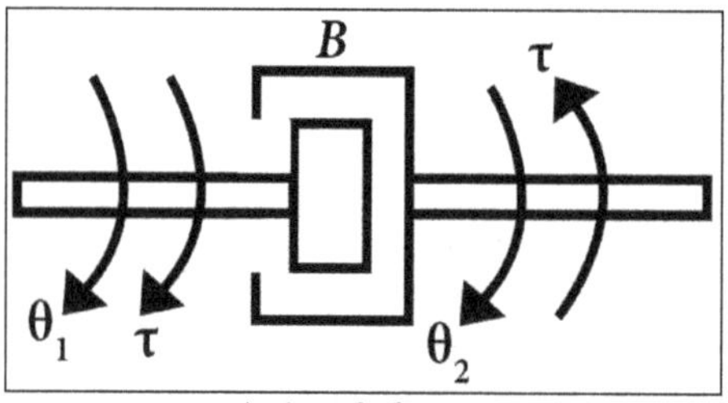

Frictional element.

Let,

τ = Applied torque.

τ_b = Opposing torque due to friction.

Here,

$$\tau_b = \alpha \, B \, d/dt \, (\theta_1 - \theta_2)$$

By Newton's law,

$$\tau = \tau_b = B \, d/dt \, (\theta_1 - \theta_2)$$

3. Consider an ideal elastic element, torsional spring as shown in figure, which has negligible moment of inertia and friction. Let a torque be applied on it. The torsional spring will offer an opposing torque which is proportional to the angular displacement of the body.

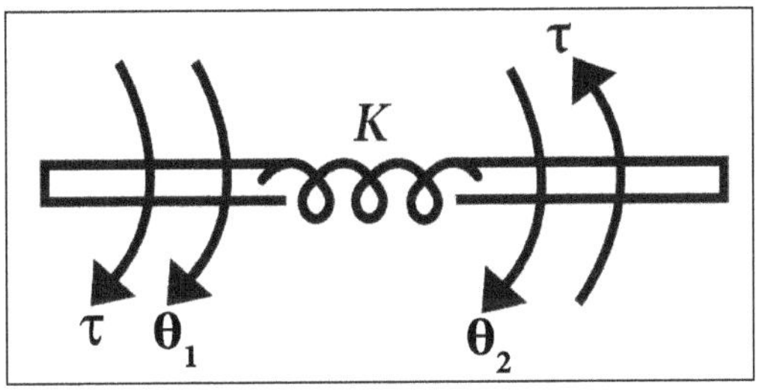

Elastic element.

Let,

τ = Applied torque.

τ_k = Opposing torque due to friction.

Here,

$$\tau_k \alpha \, K(\theta_1 - \theta_2)$$

By Newton's law,

$$\tau = \tau_k = K \, (\theta_1 - \theta_2)$$

Problems

1. Consider the mechanical system shown below. For this figure, let us identify the variables and also write the differential equation.

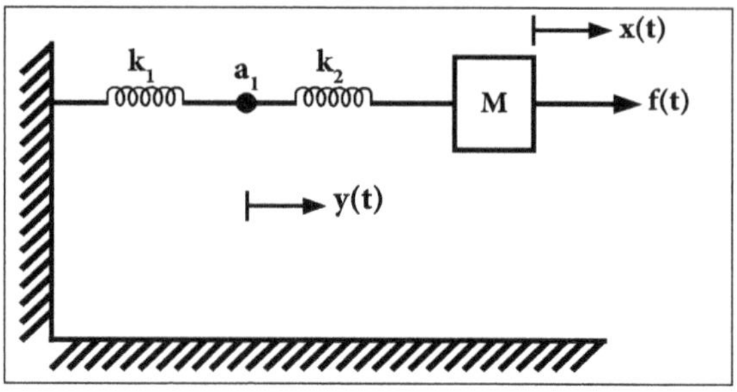

Solution:

By Newton's second law,

$$fK_1 + fK_2 = 0$$

$$K_1 y + K_2 y = 0$$

$$y(K_1 + K_2) = 0$$

Taking Laplace transform,

$$Y(s) = [K_1 + K_2] = 0$$

Consider the mass M. The displacement at mass M be x. The applied force be f.

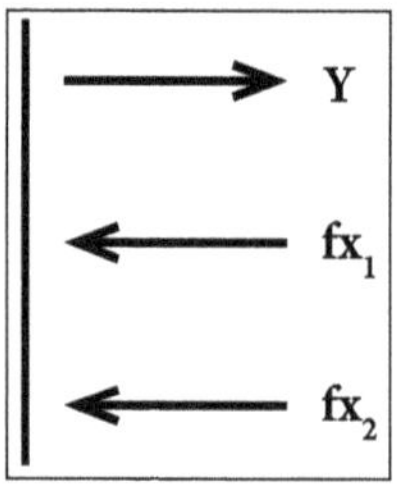

By Newton's second law,

$$f(t) = f_m + fK_2$$

$$f(t) = M\frac{d^2x}{dt^2} + K_2(x - y)$$

Taking Laplace transform on both the sides,

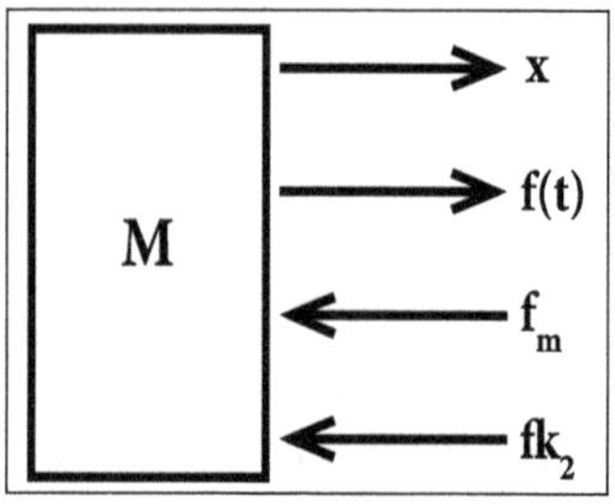

$$F(s) = Ms^2X(s) + K_2(X(s) - Y(s))$$

$$F(s) = X(s)[Ms^2 + K_2] - Y(s)[K_2]$$

2. Let us write the differential equations governing the mechanical rotational system for the below figure and also determine the torque-voltage and torque-current electrical analogous circuits. Finally, let us also verify it through the mesh and node equations.

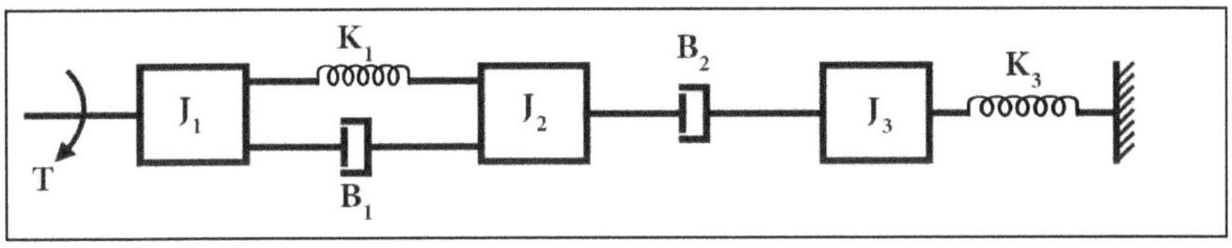

Solution:

Formula to be used:

Torque-Voltage analogous circuit,

$$T \rightarrow e(t); \omega_1 \rightarrow i_1; J_1 \rightarrow L_1; B_1 \rightarrow R_1; M_1 \rightarrow \frac{1}{C_1}$$

$$\omega_2 \rightarrow i_2\ J_2 \rightarrow L_2\ B_2 \rightarrow R_2\ K_3 = \frac{1}{C_3}$$

Torque-Current analogous circuit:

$$T > i(t); \omega_1 > V_1; J_1 > C_1$$

$$B_1 > 1/R_1; K_1 > 1/L_1$$

$$\omega_2 > V_2; J_2 > C_2; B_2 > 1/R_2; K_3 > 1/L_3$$

$$\omega_3 > V_3; J_3 > C_3$$

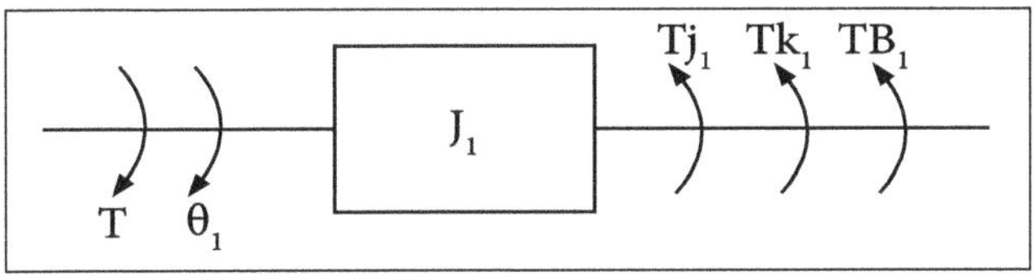

Free body diagram of J_1.

The opposing torques are,

$$T_{j1} = \frac{J_1 d^2 \theta_1}{dt_2}; T_{B1} = B_1 \cdot \frac{d(\theta_1 - \theta_2)}{dt}$$

$$T_{k1} = k_1(\theta_1 - \theta_2).$$

By Newton's second law,

$$T_{j1} + T_{B1} + T_{k1} = T$$

$$T = \frac{J_1 d^2\theta_1}{dt^2} + B_1 \frac{d}{dt}(\theta_1 - \theta_2) + K_1(\theta_1 - \theta_2) \quad ...(1)$$

The opposing torque of J_2 are $T_{j2}, T_{k1}, T_{B1}, T_{B2}$

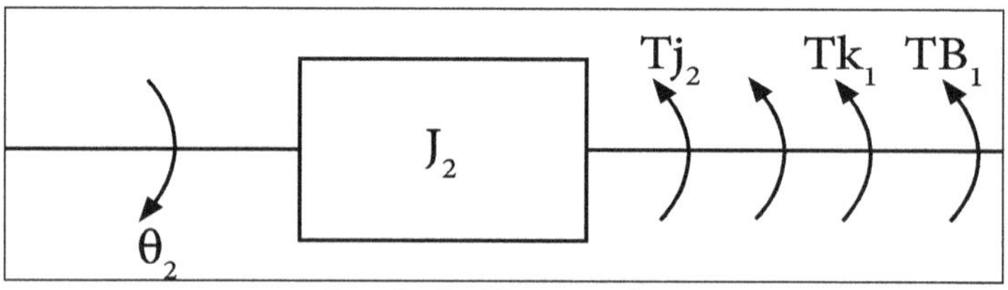

Free body diagram of J_2.

By Newton's second law,

$$Tj_2 + Tk_1 + TB_1 + TB_2 = 0$$

$$J_2 \cdot \frac{d^2\theta_2}{dt^2} + K_1 \cdot (\theta_2 - \theta_1) + B_1 \cdot \frac{d(\theta_2 - \theta_1)}{dt} + B_2 \cdot \frac{d(\theta_2 - \theta_3)}{dt} = 0 \quad ...(2)$$

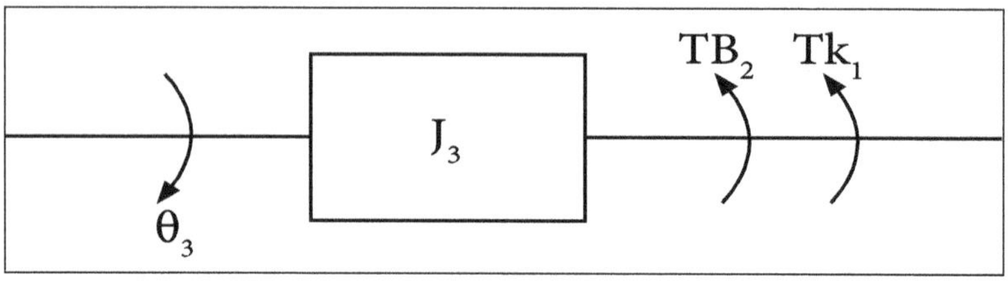

Free body diagram of J_3.

The opposing torque of J_3 are TB_2 and TK_3.

By Newton's second law,

$$TB_2 + TK_3 = 0$$

$$B_2 \cdot \frac{d(\theta_3 - \theta_2)}{dt} + K_3\theta_3 = 0 \quad ...(3)$$

$$\frac{d^2\theta}{dt^2} = \frac{d\omega}{dt}; \frac{d\theta}{dt} = \omega,$$

On replacing $\theta = \int \omega \, dt$ in (1), (2) and (3), equations will be rewritten as,

$$J_1 = \frac{d\omega_1}{dt} + B_1(\omega_1 + \omega_2) + K_1 \int (\omega_1 + \omega_2) dt = T$$

$$J_2 = \frac{d\omega_2}{dt} + B_1(\omega_2 - \omega_1) + B_2(\omega_2 - \omega_3) + K_1 \int (\omega_2 - \omega_1) = 0$$

$$J_3 \frac{d\omega_3}{dt} + B_2(\omega_3 - \omega_1) + K_3 \int \omega_3 \, dt = 0$$

Torque-Volt Age Analogous Circuit

The electrical analogous elements for elements of mechanical rotational system is expressed as:

$$T = e(t); \omega_1 \rightarrow i_1; J_1 \rightarrow L_1; B_1 \rightarrow R_1; M_1 \rightarrow \frac{1}{C_1}$$

$$\omega_2 \rightarrow i_2 \, J_2 \rightarrow L_2 \, B_2 \rightarrow R_2 \, K_2 \rightarrow \frac{1}{C_3}$$

$$\omega_3 \rightarrow i_3 J_3 \rightarrow L_3$$

By Kirchhoff's voltage law,

$$L_1 \frac{di_1}{dt} + R_1(i_1 - i_2) + \frac{1}{C_1} \int (i_1 - i_2) dt = e(t)$$

$$L_2 \frac{di_2}{dt} + R_2(i_2 - i_3) + R_1(i_2 - i_1) + \frac{1}{C_1} \int (i_2 - i_1) dt = 0$$

$$L_3 \frac{di_3}{dt} + R_2(i_3 - i_2) + \frac{1}{C_3} \int i_3 dt = 0$$

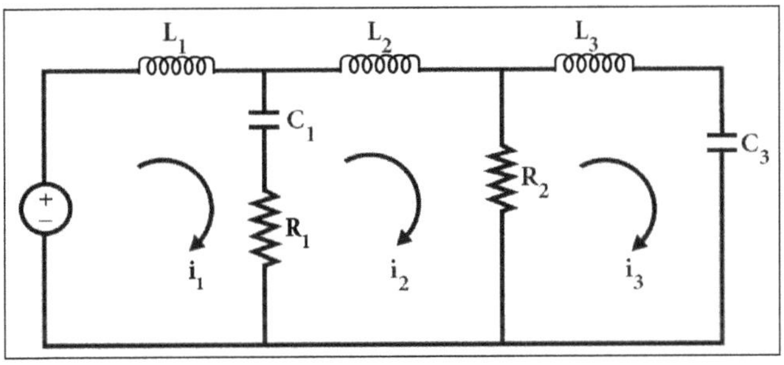

Torque-Current Analogous Circuit

The electrical analogous elements for the elements of mechanical rotational system are,

$T > i(t);\ \omega_1 > V_1\ ;\ J_1 > C_1$

$B_1 > 1/R_1\ ;\ K_1 > 1/L_1$

$\omega_2 > V_2\ ;\ J_2 > C_2\ ;\ B_2 > 1/R_2\ ;\ K_3 > 1/L_3$

$\omega_3 > V_3;\ J_3 > C_3.$

By using Kirchhoff's current law,

$$C_1 \cdot \frac{dV_1}{dt} + \frac{1}{R_1}(V_1 - V_2) + \frac{1}{L_1}\int (V_1 - V_2)dt = i(t)$$

$$C_2 \cdot \frac{dV_2}{dt} + \frac{1}{R_1}(V_2 - V_1) + \frac{1}{R_2}\int (V_2 - V_3) + \frac{1}{L_1}\int (V_2 - V_1)dt = 0$$

$$C_3 \cdot \frac{dV_3}{dt} + \frac{1}{R_2}(V_3 - V_2) + \frac{1}{L_3}\int V_3 \cdot dt = 0$$

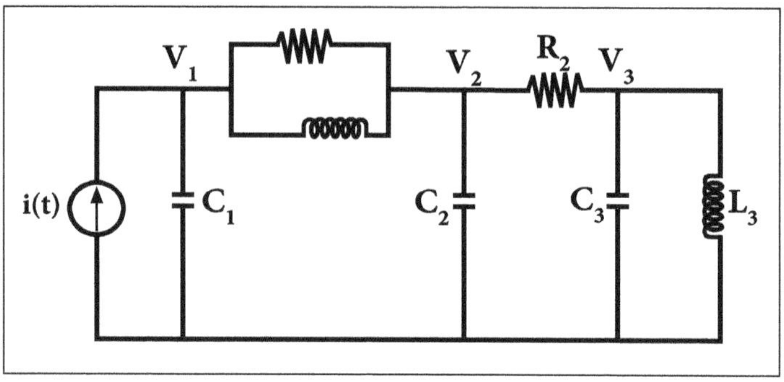

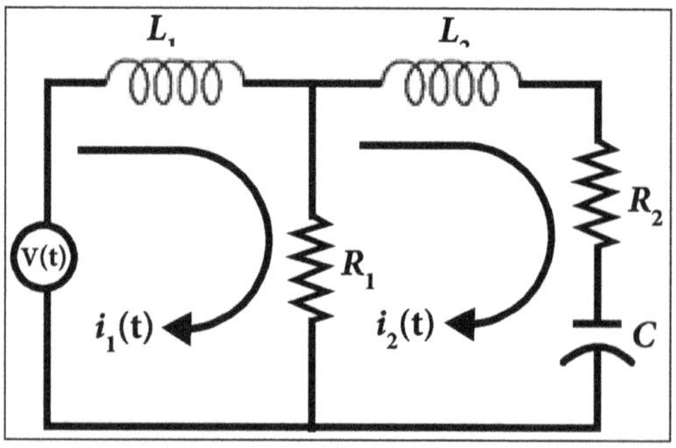

3. Let us convert the given mechanical system into force voltage and force current analogy.

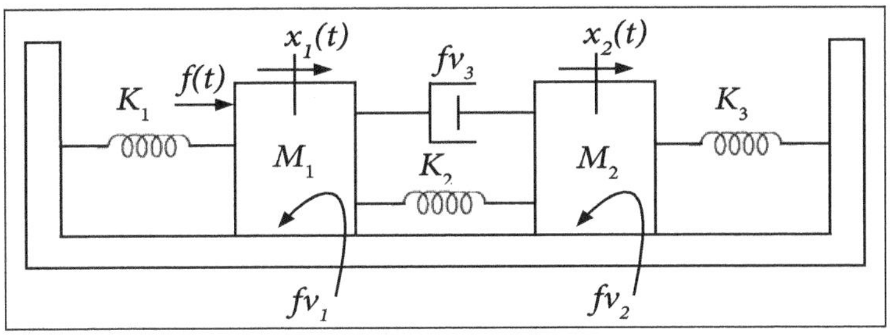

Solution:

To find:

Conversion of mechanical system into force voltage and force current analogy.

(1) Force voltage analogy of the given mechanical system:

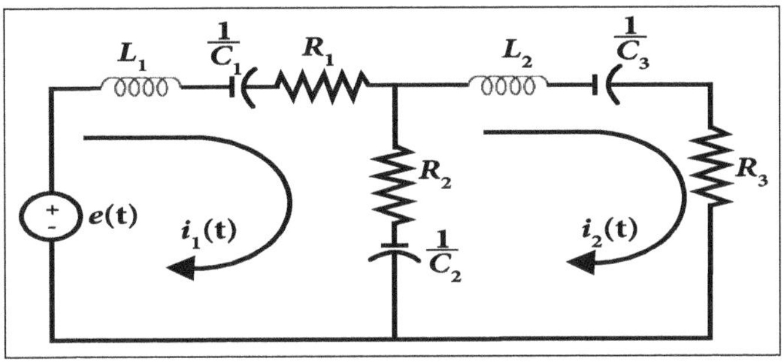

(2) Force current analogy of the given mechanical system:

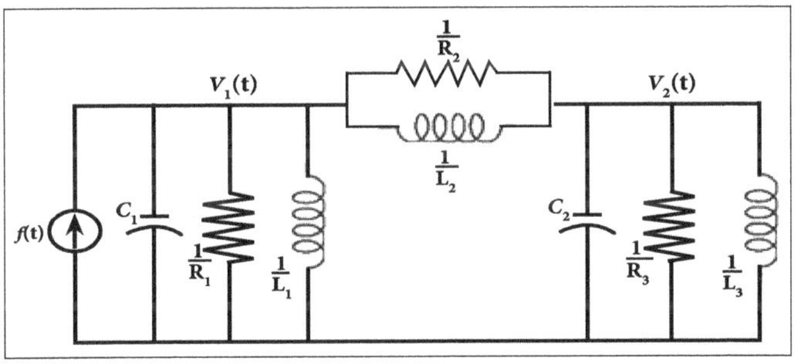

5. Let us draw the voltage and current analogy for the following mechanical system.

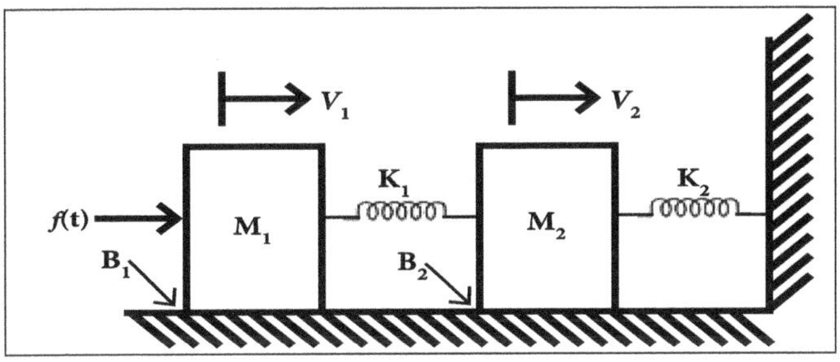

Solution:

Force Voltage analogy:

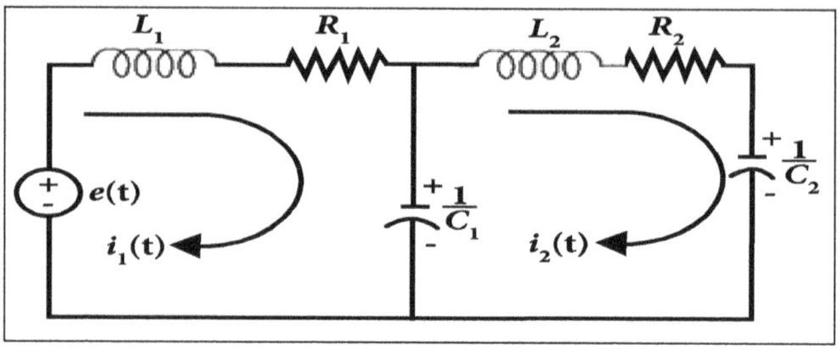

Force Current analogy:

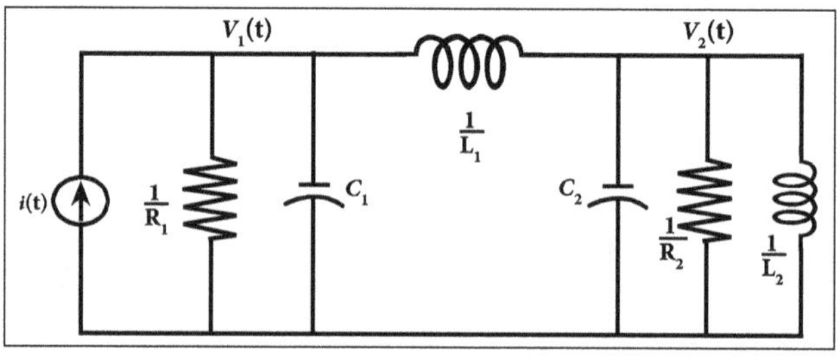

Problems Based on Rotational Mechanical System

1. Let us derive the differential equations governing the mechanical rotational system for the below figure. Also let us draw the torque-current electrical analogous circuit and verify by writing the node equations.

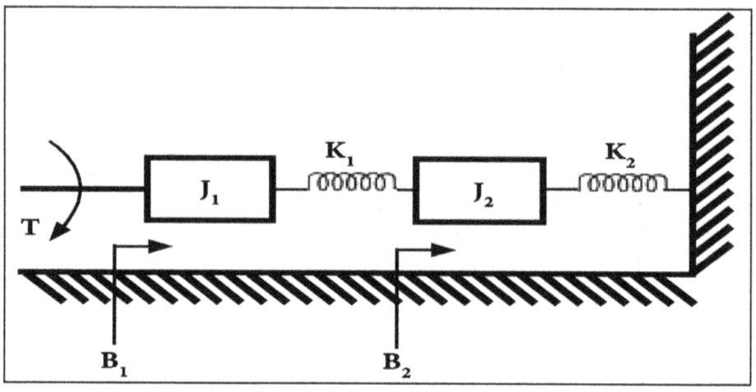

Solution:

To find: Differential equations,

Torque Current Electrical Analogous Circuit

The differential equation governing the mechanical rotational system is given by,

$$J_1 \frac{dw_1}{dt} + B_1 w_1 + K_1 \int (w_1 - w_2) dt = T$$

$$J_2 \frac{dw_2}{dt} + B_2 w_2 + K_2 \int w_2 \, dt + K_1 \int (w_1 - w_2) dt = 0$$

(We know that the angular frequency, $w = d\theta / dt$)

Mechanical rotational system	Torque-Current analogous	Torque-Voltage analogous
Torque (T)	Current i(t)	Voltage v(t)
Moment of Inertia (J)	Capacitance (C)	Inductance (L)
Displacement (θ)	F lux (φ)	Charge (q)
Viscous friction coefficient (B)	Conductance (1/R)	Resistance (R)
Angular velocity (w)	Voltage v(t)	Current i(t)
Stiffness co - efficient (K)	Reciprocal of inductance (1/L)	Reciprocal of capacitance (1/C)

For the given Torque-Current analogous circuit, (we know that $v = d\phi / dt$)

The node equation is given by,

$$C_1 \frac{dv_1}{dt} + \frac{1}{R_1} v_1 + \frac{1}{L_1} \int (v_1 - v_2) dt = i(t)$$

$$C_2 \frac{dv_2}{dt} + \frac{1}{R_1} v_1 + \frac{1}{L_2} \int v_2 \, dt + \frac{1}{L_1} \int (v_2 - v_1) dt = 0$$

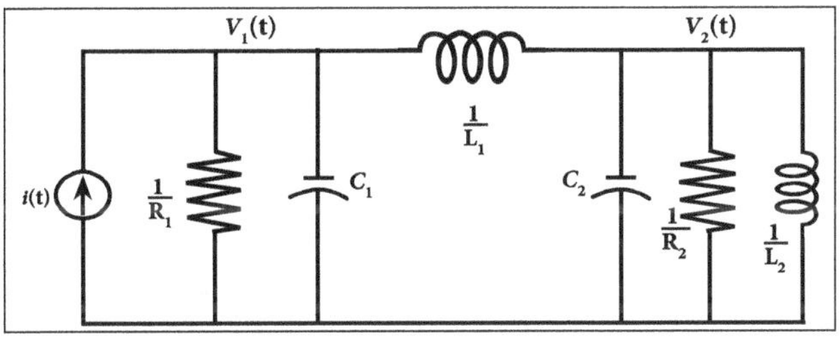

2. Let us draw the torque-voltage and torque-current electrical analogous circuit for the below mechanical system.

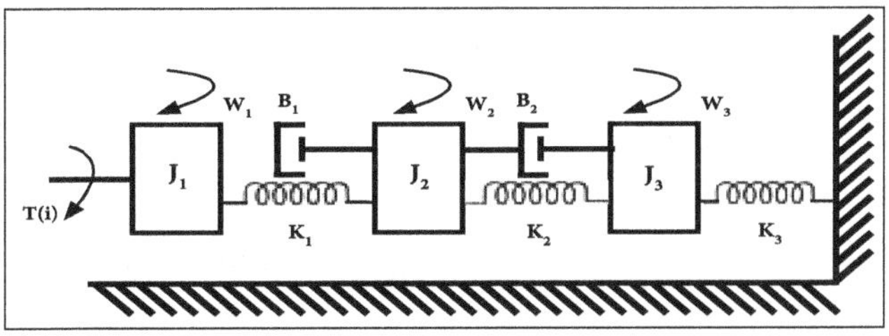

Solution:

To find:

Torque voltage and torque current electrical analogous circuit.

Torque-Voltage analogous circuit:

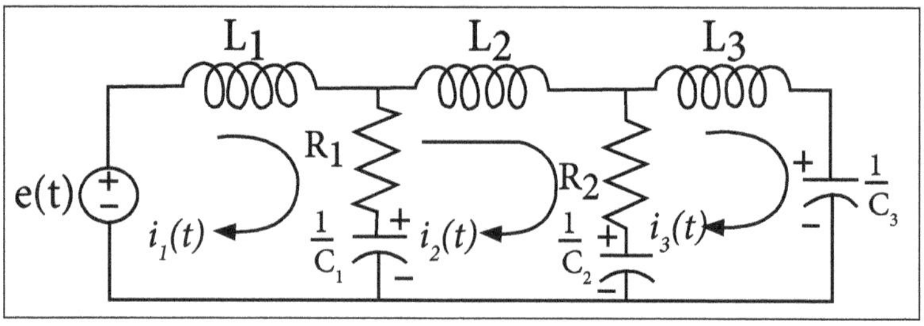

Torque-Current analogous circuit:

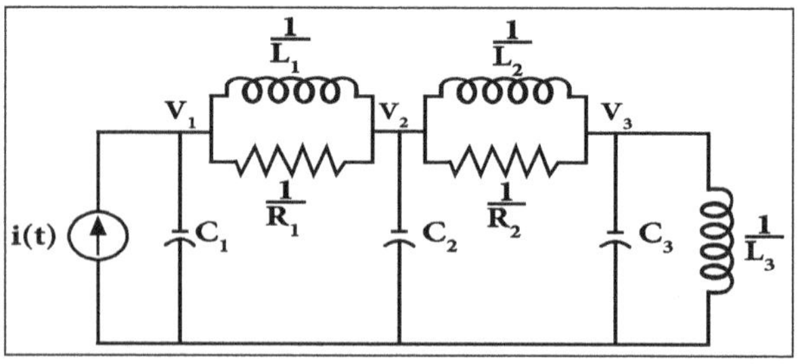

3. Let us obtain the torque-voltage electrical analogous circuit for the following mechanical system shown.

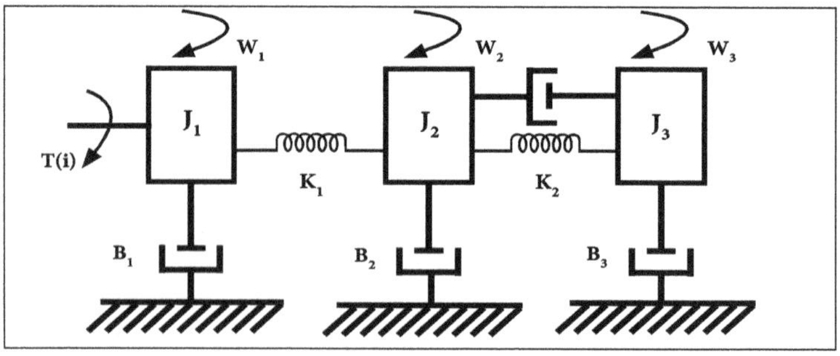

Solution:

To find:

Torque voltage electrical analogous circuit.

Torque-Voltage analogous circuit:

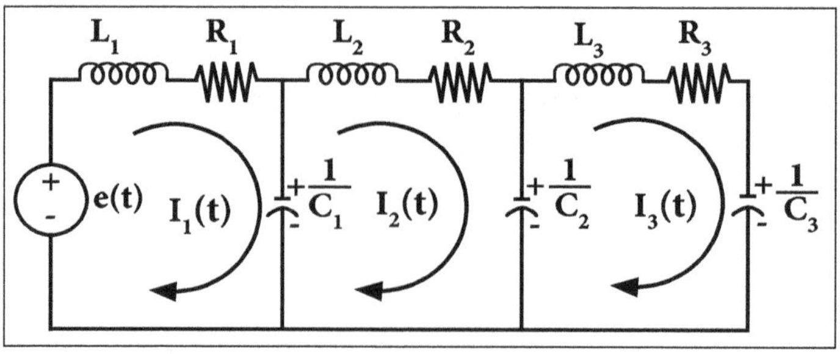

1.6 Transfer Function of DC Servo Motor

Transfer Function of Armature Controlled DC Motor

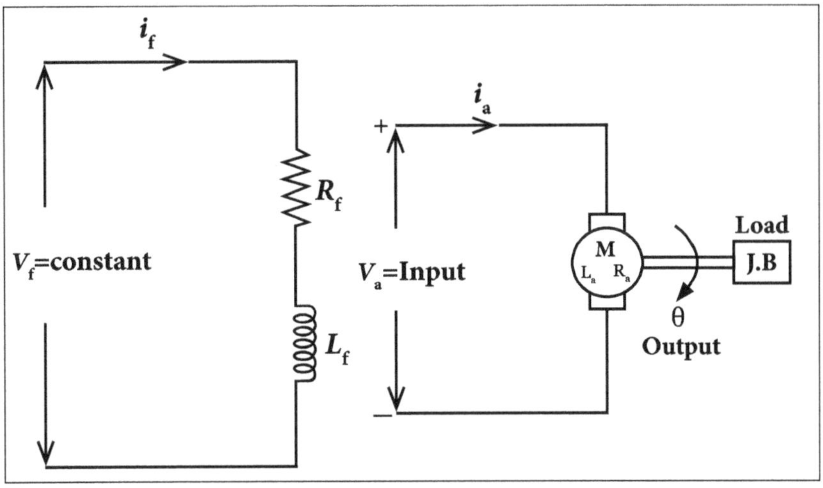

Armature controlled DC motor.

The armature controlled DC motor speed control system is shown in the above figure.

Let,

Ra = Armature resistance, Ω

L_a = Armature inductance, H

i_a = Armature current, A

V_a = Armature voltage, V

e_b = Back emf

K_t = Torque constant, N-m/A.

T = Torque developed by motor, N-m.

Q = Angular displacement of shaft, rad

J = Moment of inertia of motor and load, kg-m /rad.

B = Frictional co-efficient of motor and load, N-m/ (rad/sec).

K_b = Back emf constant, V/ (rad/sec).

The equivalent circuit of armature is shown in the below figure:

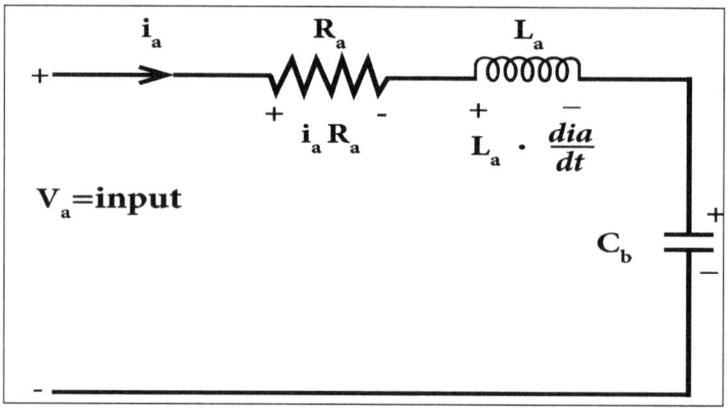

Equivalent Circuit.

By KVL, we can write,

$$i_a R_a + L_a \cdot \frac{dia}{dt} + e_b = V_a. \qquad ...(1)$$

Torque of the DC motor is directly proportional to the product of flux and current. Since flux is constant in this system, the torque is proportional to i_a alone

$$T \alpha \, i_a$$

$\therefore$ Torque, $T = K_t \cdot i_a$

Where,

K_t = Torque constant.

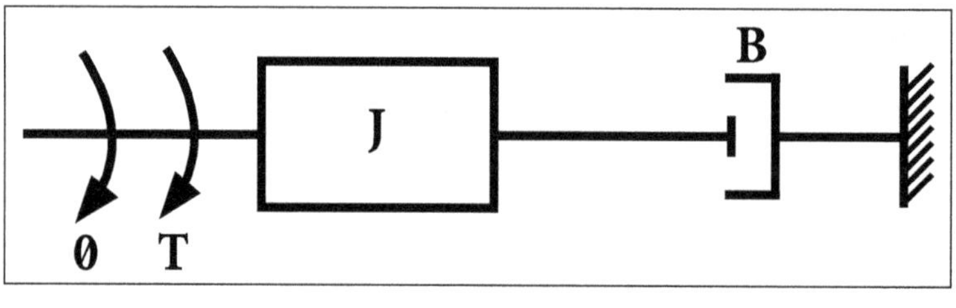

The mechanical system of the motor is shown. The differential equation governing the mechanical system of motor is given by,

$$J \cdot \frac{d^2\theta}{dt^2} + B \cdot \frac{d\theta}{dt} = T \qquad ...(2)$$

The back emf of DC machine is proportional to speed of shaft.

$$\therefore \; e_b \propto \frac{d\theta}{dt};$$

Back emf,

$$C_b = K_b \cdot \frac{d\theta}{dt} \qquad ...(3)$$

The differential equations governing the armature controlled DC motor speed control system are:

$$i_a R_a + L_a \cdot \frac{di_a}{dt} + e_b = V_a\,;\, J\frac{d^2\theta}{dt^2} + B\frac{d\theta}{dt} = T$$

$$T = Kt\ .i_a\ e_b = K_b \cdot \frac{d\theta}{dt}.$$

On taking L.T of the system differential equations with zero initial conditions.

We get,

$$I_a(s)R_a + L_a(s)I_a(s) + E_b(s) = V_a(s) \qquad ...(4)$$

$$T(s) = K_t\ I_a\ (s) \qquad ...(5)$$

$$J_s^2\theta + B_s\theta(s) = T(s) \qquad ...(6)$$

$$E_b(s) = K_{bs}\ \theta(s) \qquad ...(7)$$

On equating (5) and (6), we get,

$$K_t I_a(s) = (J_s^2 + B_s)\,\theta(s) \Rightarrow I_a(s) = \frac{(J_s^2 + B_s)}{K_t}\theta(s) \qquad ...(8)$$

Equation (4) can be written as:

$$(R_a + sL_a)\ I_a(s) + E_b(s) = V_a(s) \qquad ...(9)$$

Substitute for E_b (s), I_a (s) and I_s (s) in equations (7), (8) and (9) respectively,

$$(R_a + sL_a)\left(\frac{J_s^2 + B_s}{K_t}\right)\theta(s) + K_b\, s\theta(s) = V_a(s)$$

$$\left[\frac{(R_a + sL_a)(J_s^2 + B_s) + K_b \cdot K_t \cdot S}{K_t}\right]\theta(s) = V_a(s).$$

$$\frac{\theta(s)}{V_a(s)} = \frac{K_t}{(R_a + sL_a)(J_s^2 + B_s) + K_b K_t s}$$

$$= \frac{K_t}{s\left[JL_a S^2 + (JR_a + BL_a)S + (BR_a + K_b K_t)\right]}$$

$$\Rightarrow \frac{K_t / JL_a}{S\left[S^2 + \left[\frac{JR_a + BL_a}{JL_a}\right]S + \left[\frac{BR_a + K_b K_t}{IL_a}\right]\right]}$$

$$\Rightarrow \frac{K_t / R_a B}{s\left[(1 + sT_a)(1 + sT_m) + \frac{K_b K_t}{R_a B}\right]}$$

Where,

$L_a / R_a = T_a =$ Electrical time constant.

$J / B = T_m =$ Mechanical time constant.

1.6.1 AC Servo Motor

Transfer Function of AC Servo Motor

Various approximations to derive transfer function are as follows:

- A servomotor rarely operates at high speeds. Hence for a given value of control voltage, T α N characteristics is perfectly linear.
- In order that T α N characteristics are directly proportional to the voltage applied to its control phase, we assume T α N characteristics are straight lines and equally spaced.

Torque at any speed 'N' is given by,

$$T_m = K_m E_{2t} + m\frac{d\theta_m}{dt}$$

Where,

$\frac{d\theta_m}{dt}$ is speed of motor.

If load consists inertia J_m and friction B_m we can write,

$$T_m(s) = J_m\ s_2\ \theta_m + B_m\ s\theta_m$$

Now Laplace transform of the above equation is given by,

$$Tm\ (s) = K_{tm}\ E_2(s) + m_s\theta_m\ (s)$$

On equating equations we get,

$$\therefore\ Kt_mE_2\ (s) + m_s\ \theta_m\ (s) = J_m\ s2\ \theta_m\ + B_m\ s\theta_m\ (s)2\theta$$

$$\therefore\ \frac{\theta_m(s)}{E_2(s)} = \frac{K_{tm}}{s(sJ_m - m + B_m)} = \frac{K_{tm}}{s(B_m - m)\left[1 + \frac{sJ_m}{(B_m - m)}\right]}$$

$$\frac{\theta_m(s)}{E_2(s)} = \frac{K_m}{s(1+\tau_m s)}$$

Where,

$$K_m = \frac{K_{tm}}{B_m - m}$$

And $\tau_m = \dfrac{J_m}{B_m - m}$

As slope is negative in the above equation, $[B_m - m]$ indicates that the total friction increases due to the value of m.

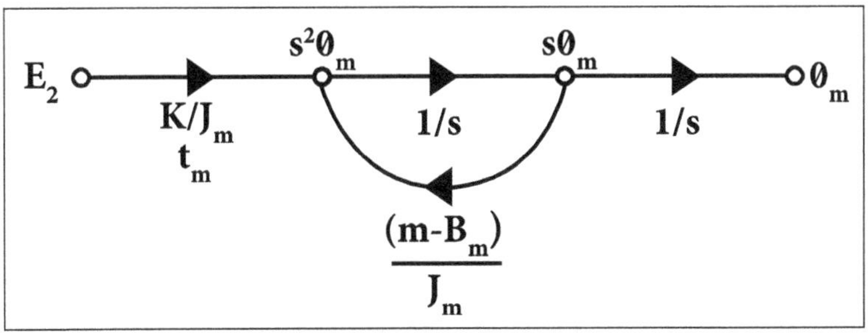

Signal flow graph of AC servomotor.

As it adds more friction, the damping improves the stability of the motor. This is called Internal Electrical Damping of 2ph AC servomotor.

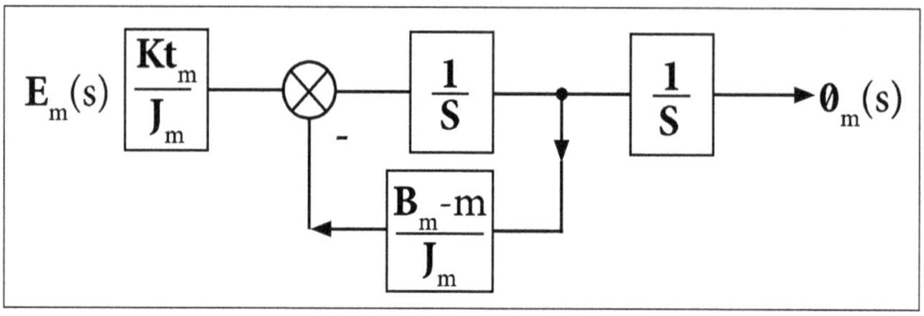

Block diagram of AC servomotor.

1.6.2 Synchronal, Transmitter and Receiver

Synchs are simply variable transformers. It consists of two electro-mechanical devices. Each synchro contains:

- A rotor, which is similar in appearance to the armature in a motor.
- A stator, which corresponds to the field in a motor.

The stator consists of a balanced three phase winding and is star connected. The rotor is of dumb-bell type construction and is wound with a coil to produce a magnetic field. When no voltage is applied to the winding of the rotor, a magnetic field is produced. The coils in the stator link with this sinusoidal distributed magnetic flux.

Voltages are induced in the three coils due to transformer action. Three voltages are in time phase with each other and the rotor voltage. The magnitudes of the voltages are proportional to the cosine of the angle between the rotor position and its respective coil axis.

A synchro resembles a small electrical motor in size and appearance and it operates like a variable transformer.

Synchros are variable transformers. Each synchro contains a rotor, similar in appearance to the armature in the motor, and a stator, which corresponds to the field in the motor.

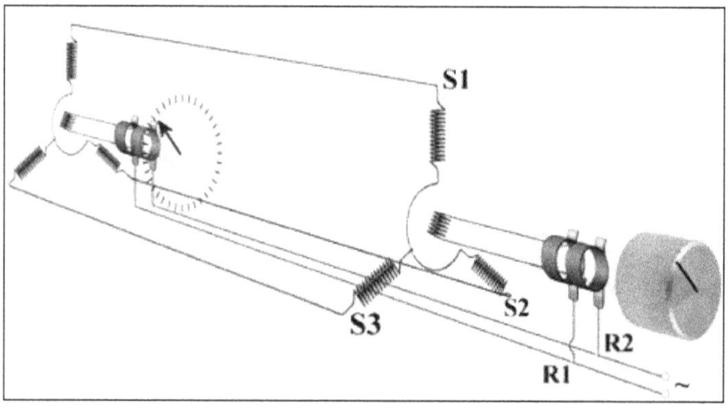

Principle of torque synchro measurement.

Torque-Synchro Systems

Torque-synchro systems are classified as "torque" because they are mainly concerned with the torque or turning force required to move the light loads such as pointers, dials or similar indicators. The positioning of these devices requires a relatively low amount of torque. The synchro stator is composed of three Y- connected or Δ-connected windings spaced 120° apart. Both stationary and rotating coils are connected to the same supply voltage.

The torque developed in the synchro receiver results from the tendency of two electromagnets to align themselves. Since the rotor can be turned and the stator usually cannot, the stator must exert a force tending to pull the rotor into a position where the primary and the secondary magnetic fields are in line.

The strength of the magnetic field produced by the stator determines the torque. The field strength depends on the current through the stator coils. As the current through the stator is increased, the field strength increases and more torque is developed.

Control Synchros

Control synchros are used in the systems which are designed to move heavy loads such as gun directors, radar antennas, and missile launchers. A positioning servo system using a control synchro system consisting of a servo amplifier, a servomotor, a control transmitter and a control transformer.

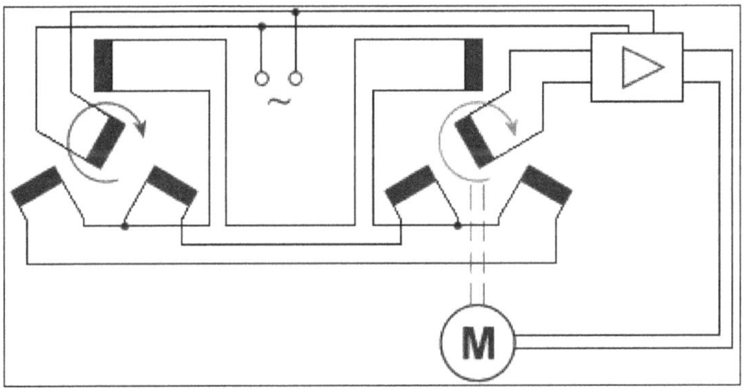

Typical electromechanical follow-up servo system.

The error signal is amplified by the servo amplifier and applied to the servomotor. The servomotor turns the load, and through a mechanical linkage called "response", also turns the rotor of the control transformer.

Servomotor t urns the rotor of the control transformer so that it is once again in correspondence with the rotor of the servomotor, the error signal drops to zero volts and the system comes to a stop.

If the data to be transmitted covers only a small range of values, a single-speed system is normally accurate. However, in the applications where as the data covers a wide range of values and the accuracy of the system is most important, the 1-speed system is not adequate enough and must be replaced by more suitable system.

Increasing the speed of a single-speed system from 1-speed up to 36-speed provides greater accuracy. The basic dual-speed synchro system consists of two transmitters and two receivers. The two speed of this system are often referred to as high and low, fast and slow, or fine and coarse.

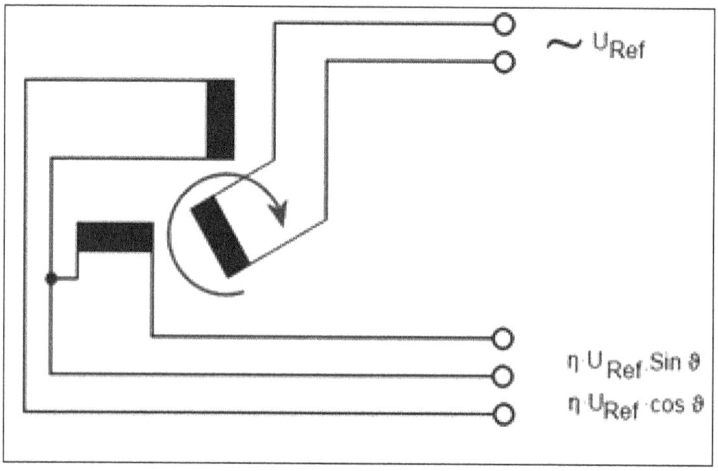

Resolver

Other applications need resolvers with two right-angle components. Physically, resolvers are similar to synchros and are used to perform the mathematical computations electrically. They are rotary electro mechanical devices that provide s the outputs that are trigonometric functions of their inputs. They are used extensively in the radar sets, analog computers, direction and target designation equipment.

Transmitter

The synchro transmitter (CG) consists of a single -phase, salient-pole rotor and a three-phase, Y-connected stator. The primary or input winding is the rotor whereas the stator is the secondary or output element.

Rotor is excited through a pair of slip rings with an AC voltage. The field produced by the input voltage induces a voltage into each of the stator phases. The magnitude of the induced phase-voltage depends on the angle between the rotor field and the resultant axis of the coils forming that stator phase. Since the axes of three stator phases are 120° apart, the magnitudes of stator output voltages can be written as follows:

$$V_{s1\text{-}3} = k\,V_{r2\text{-}1} \sin\theta$$

$$V_{s3\text{-}2} = k\,V_{r2\text{-}1} \sin(\theta+120)$$

$$V_{s2\text{-}1} = k\,V_{r2\text{-}1} \sin(\theta+240)$$

Where,

k is the maximum coupling transformation ratio (TR).

TR is further defined as,

$$T_R = \frac{V_{out}(max)}{V_{in}}$$

And is a scalar quantity:

- θ is the rotor position angle.
- V_{s1-3} is the voltage from the S_1 terminal to the S_3 terminal, and all other voltages.

These stator voltages are either approximately in the time-phase or 180° out-of-time-phase with the applied voltage. The amount by which the output voltages differ from the exact 0° or 180° time phase relationship with the input voltage is termed as synchro phase shift.

For a synchro operated at 400Hz which is working into an open circuit, the output voltage always lead the input voltage by few degrees. From the above transmitter equations, it can be readily seen that nowhere over the entire 360° rotation of the rotor will have same set of stator voltages appear.

The transmitter thus supplies the information about the rot or position angle as a set of three output voltages. To make use of this information, however, it is necessary to find the instrument which will measure the magnitude of these voltages, examine their time-phase relationships and return them to their origin al form which is a shaft position. Such an instrument is the synchro receiver (CR). These two units such as the transmitter and the receiver form the most basic synchro system.

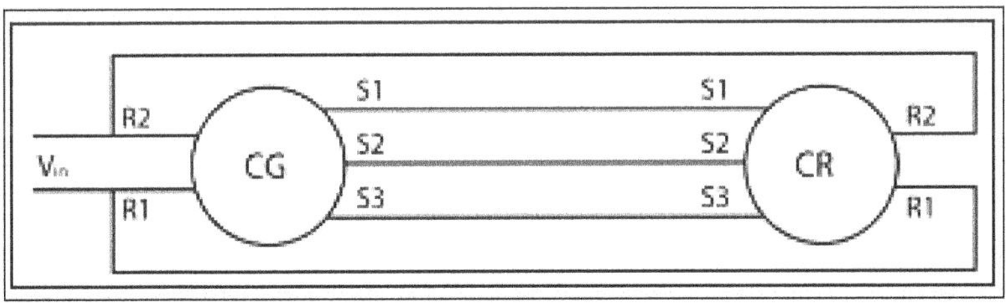

Synchro system.

Receiver

In construction, the receiver is electrically identical to the transmitter. The output voltages vary with rotor position in the identical manner as that given for the transmitter. In use, receiver is connected back - to-back with a transmitter i.e., like-numbered terminals are connected together and rotors are excited in parallel.

At the instant, the system is energized, if the rotors of each unit are not at the exact same angle relative to the stator phases, voltage differences exists across each pair of the stator windings causing current to flow in both the stators. This stator current produces a torque on each rotor.

Since the CG rotor is constrained from turning, the resultant torque acts on the CR rotor in such a direction as to align itself with transmitter. When alignment occurs, the voltages at each stator terminal are equal and opposite and thus no current flows. Perfect synchronization is never achieved because of the internal friction of receiver. To minimize this error, receiver is designed to have maximum starting friction of 2700 mg-mm.

Turning the transmitter rotor from the equilibrium position will again exert a force on the receiver rotor. As soon as this developed force exceeds the internal friction of the receiver, CR will track CG to its new position. The torque developed on the receiver shaft is proportional to the angle between the two rotors and it is usually expressed in mg-mm/deg.

Receivers are constructed to minimize the oscillations and overshoot or spinning when the rotor is turning to a new position. The time required for the rotor to reach and stabilize at its new rest position is termed as damping or synchronizing time. This time varies with the inertia of the load, size of the receiver and system torque. By special receiver construction, the damping time can be reduced if required by system considerations.

The CG-CR system is used to transmit the angular information from one point to another without mechanical linkages. The standard transmission accuracy for such a system is 30 arc minutes. Information can be sent to more than one location by paralleling more than one receiver across the transmitter. The more receivers used, however, the less accurate the system and larger the power draw from the source.

1.7 Block Diagram Algebra

The block diagram of a system is a pictorial representation of the functions performed by each component of the system and shows the flow of signals.

The basic elements of block diagram are as follows:

- Block.
- Branch point.
- Summing point.

Block

The block is the symbolic representation of transfer function of that element. The complete control system can be represented with a required number of interconnection of such blocks.

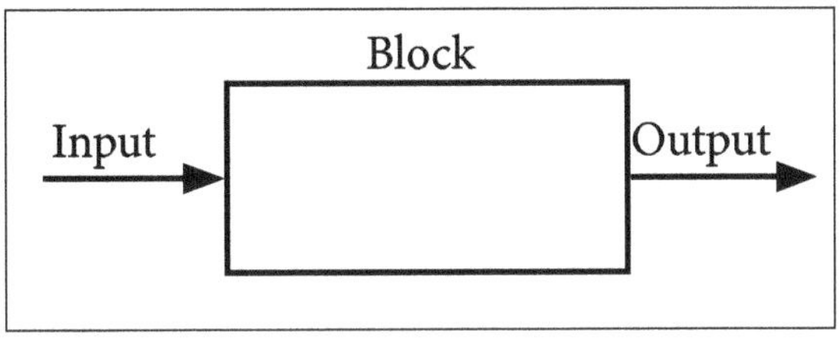

Block

Summing Point

Different input signals are applied to same block. Here, resultant input signal is the summation of all input signals applied. Summation of input signals is represented by a point called summing point which is shown in the below figure by crossed circle.

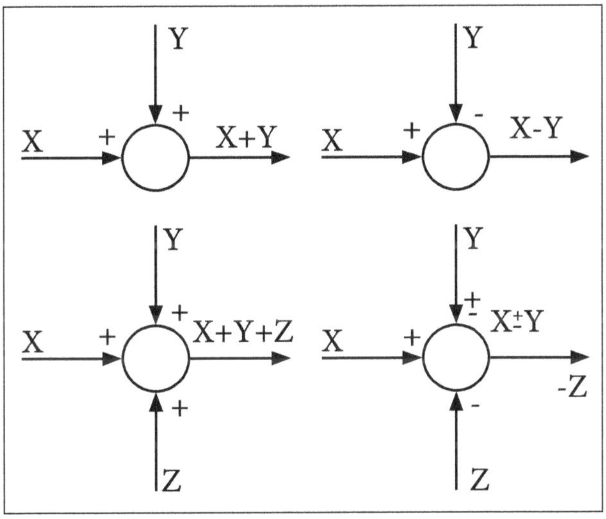

Summing point.

Block Diagram Rules

1. Cascaded blocks:

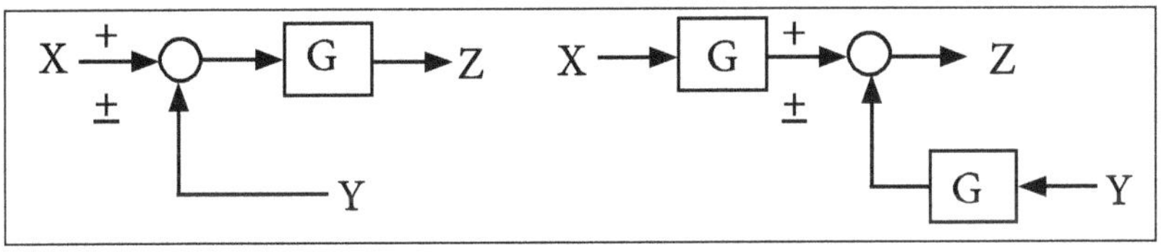

2. Moving a summer beyond the block:

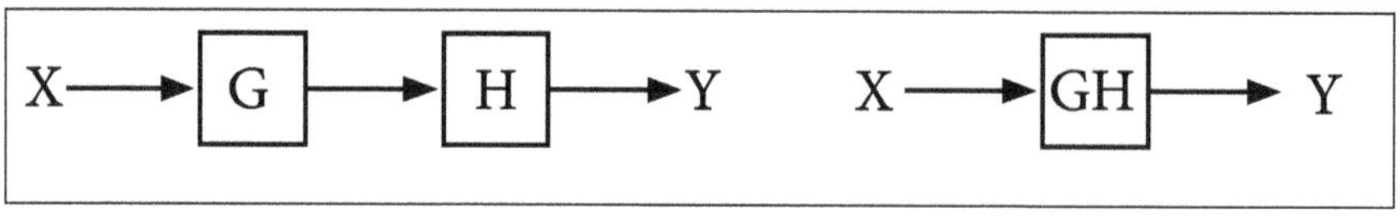

3. Moving a summer ahead of block:

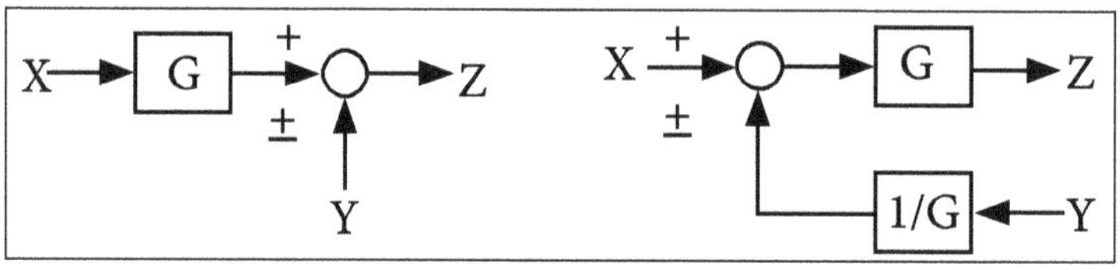

4. Moving a pick-off ahead of block:

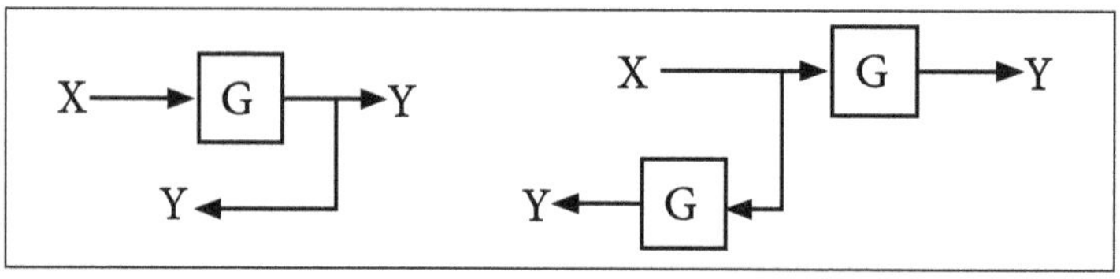

5. Moving a pick-off behind a block:

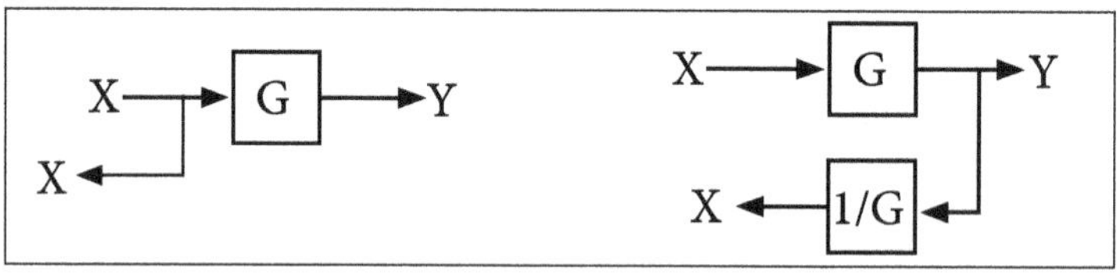

6. Eliminating a feedback loop:

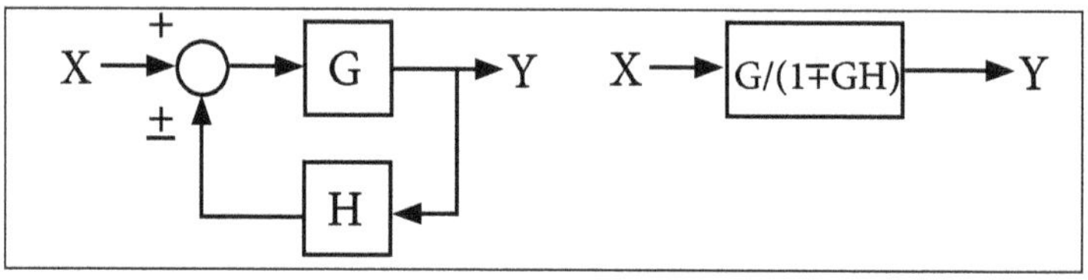

7. Cascaded Subsystems:

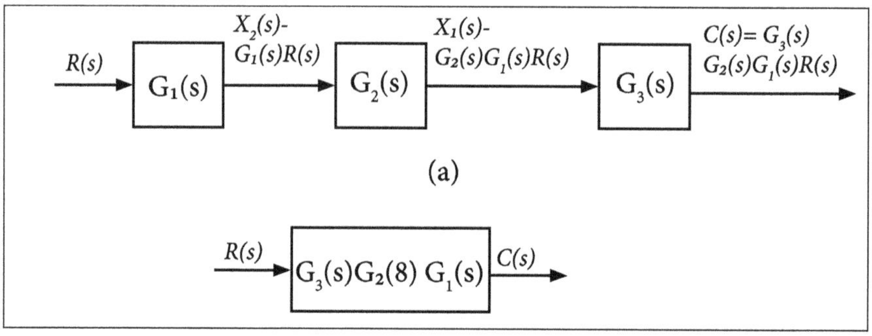

8. Parallel Subsystems:

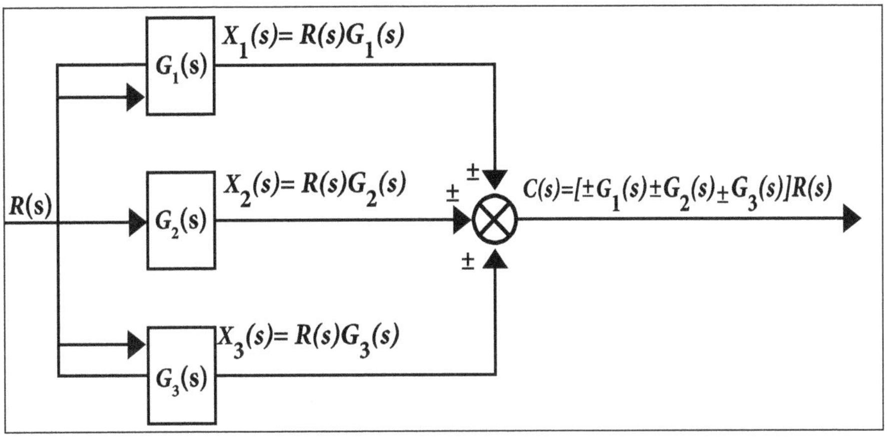

Problems

1. Let us calculate the closed-loop transfer function C/R using block diagram reduction technique whose block diagram is shown below:

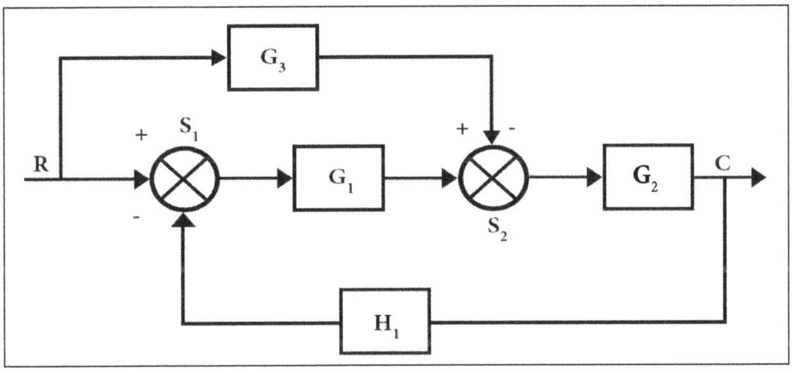

Solution:

(i) Shift the summing point S_1 after block G_1:

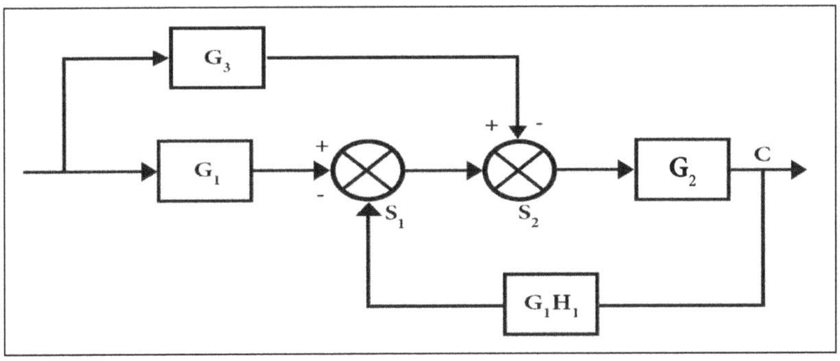

(ii) Interchanging S_1 and S_2:

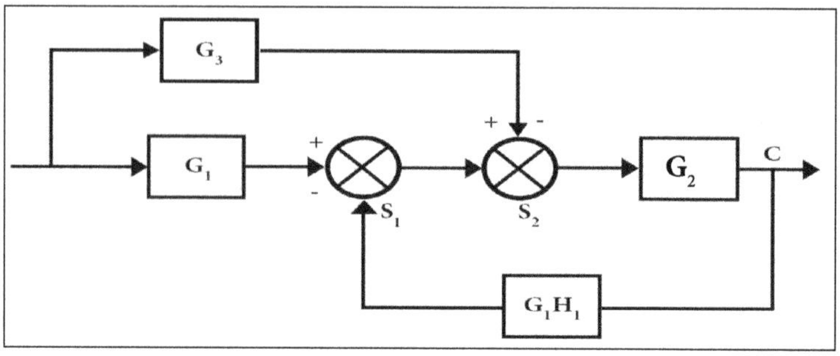

(iii) Eliminate S_1 point and S_2:

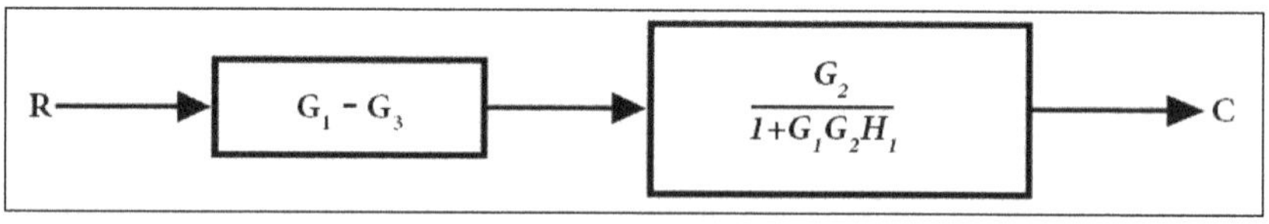

$$\therefore \quad \text{Transfer function}, \frac{C(s)}{R(s)} = \frac{(G_1 - G_3)G_2}{1 + G_1 G_2 H_1}$$

2. Using block diagram reduction method, let us determine the output of the below system.

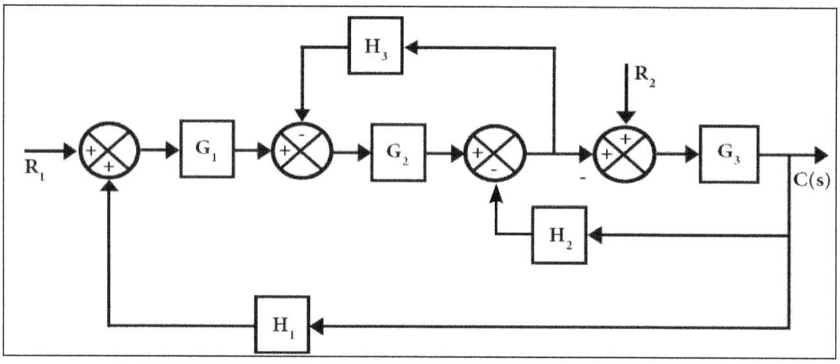

Solution:

Transfer function of the above block diagram is given by,

The Transfer function becomes $= \frac{C}{R_1} + \frac{C}{R_2}$

$$\frac{C}{R_1} = \frac{G_1 G_2 G_3}{1 + H_1 G_1 G_2 G_3 + G_3 H_2 + G_2 H_3}$$

$$\frac{C}{R_2} = \frac{G_3 (1 + G_2 H_3)}{1 + H_1 G_1 G_2 G_3 + G_3 H_2 + G_2 H_3}$$

Step 1: Shift the takeoff point before the G_4 block to a point after the block.

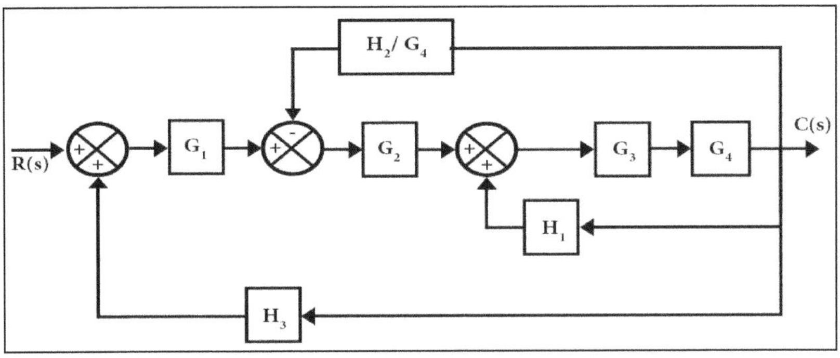

Step 2: Combine cascade blocks of G_3, G_4 and eliminate the closed loops formed by G_3 G_4 and H_1.

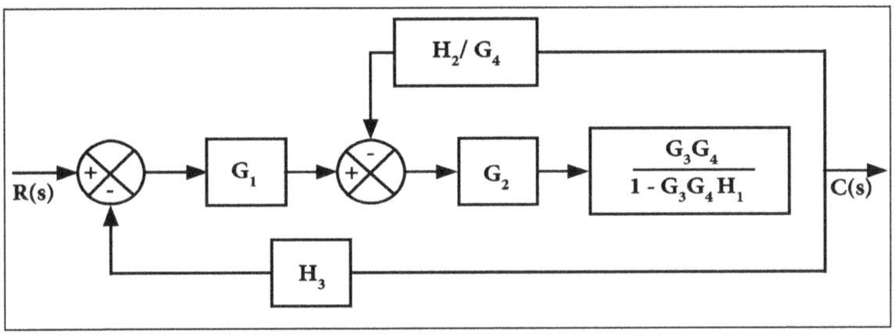

Step 3: Combine the blocks in cascade with G_2 and $\frac{G_3 G_4}{1-G_3 G_4 H_1}$. Then eliminate the closed loop formed by them with H_2/G_4.

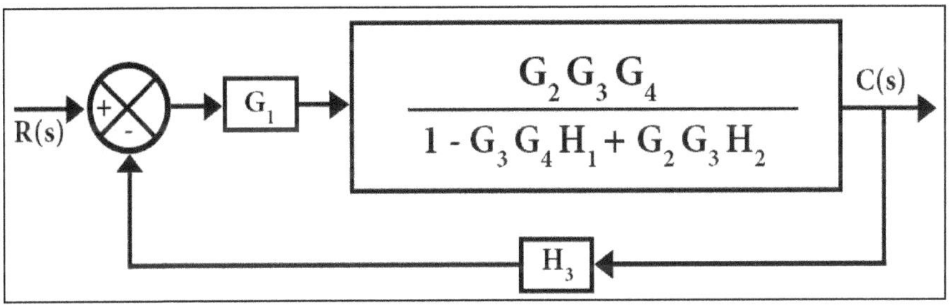

Step 4: Combine the blocks in cased with G_1 and $\frac{G_2 G_3 G_4}{1-G_3 G_4 H_1 + G_2 G_3 H_2}$.

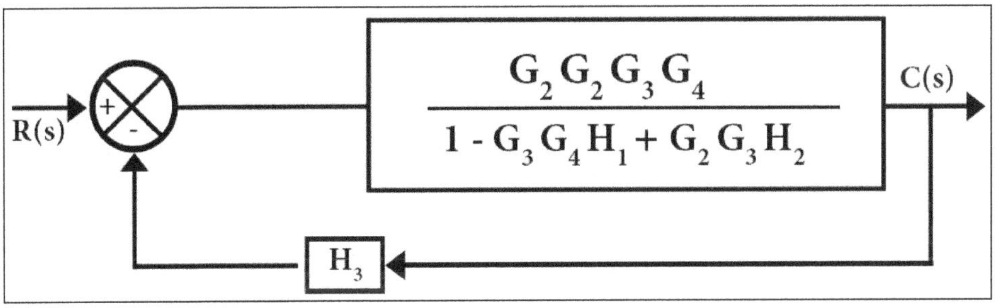

Step 5: Eliminate the feedback loop formed by H_3. The Transfer function of the given block diagram is given by,

$$\frac{C}{R} = \frac{G_1 G_2 G_3 G_4}{1 - G_3 G_4 H_1 + G_2 G_3 H_2 + G_1 G_2 G_3 G_4 H_3}$$

3. Let us calculate C(s)/R(s) for the below figure by reducing the block.

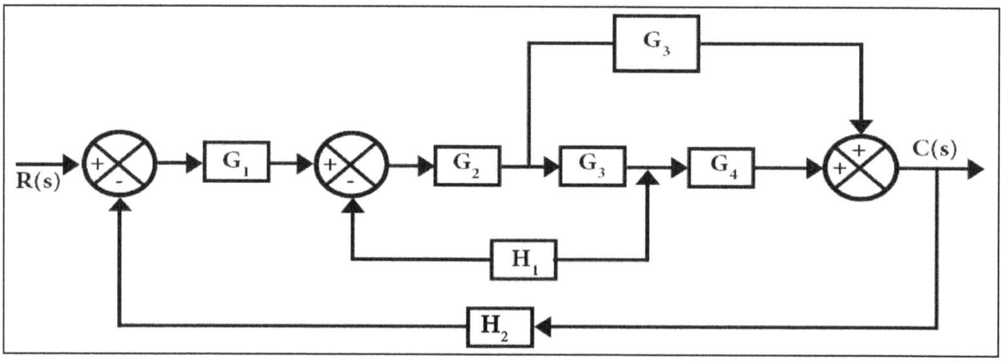

Solution:

Step 1: Shift the takeoff point after the block G_3 to before G_3.

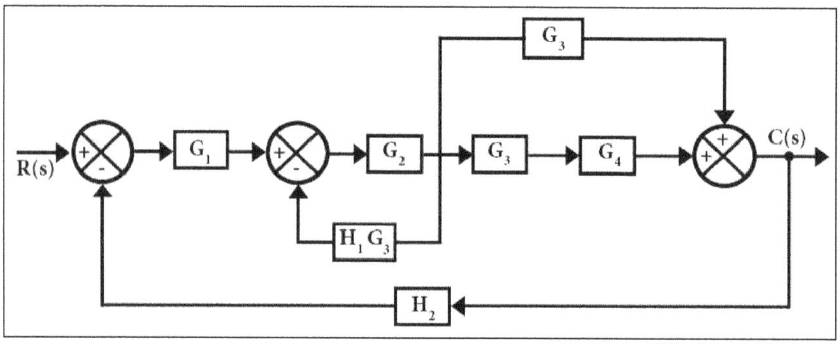

Step 2: Combine the cascade blocks G_3, G_4 and eliminate the parallel blocks.

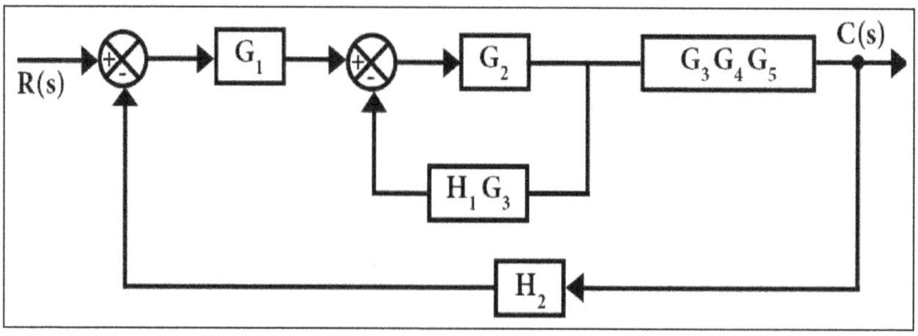

Step 3: Eliminate the feedback loop formed by G_2 and $H_1 G_3$.

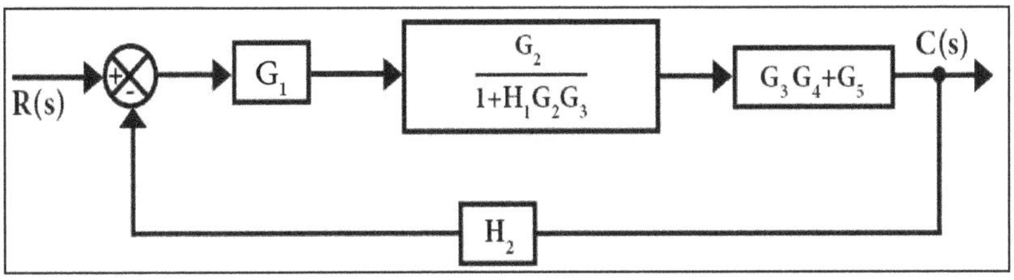

Step 4: Combine the cascade blocks.

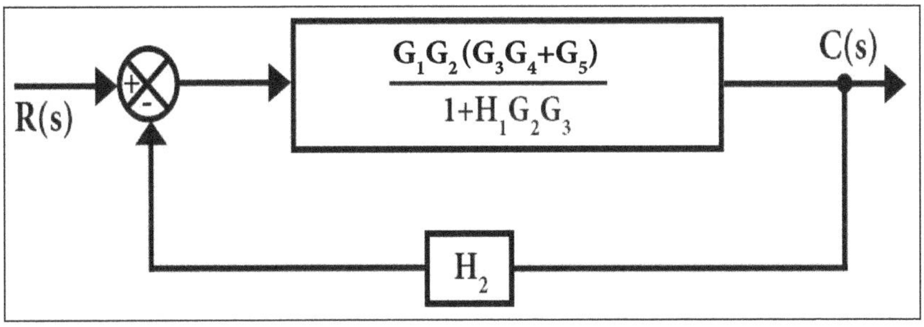

Step 5: Eliminate the feedback loop.

$$\frac{C(s)}{R(s)}=\frac{G_1G_2(G_3G_4+G_5)/1+H_1G_2G_3}{1+H_2\times\dfrac{G_1G_2(G_3G_4+G_5)}{1+H_1G_2G_3}}$$

The transfer function becomes,

$$\frac{C(s)}{R(s)}=\frac{G_1G_2(G_3G_4+G_5)}{1+H_1G_2G_3+\left[G_1G_2(G_3G_4+G_5)\right]H_2}$$

4. By Reducing the block diagram for the below figure, let us obtain its closed loop transfer function C(S)/R(S).

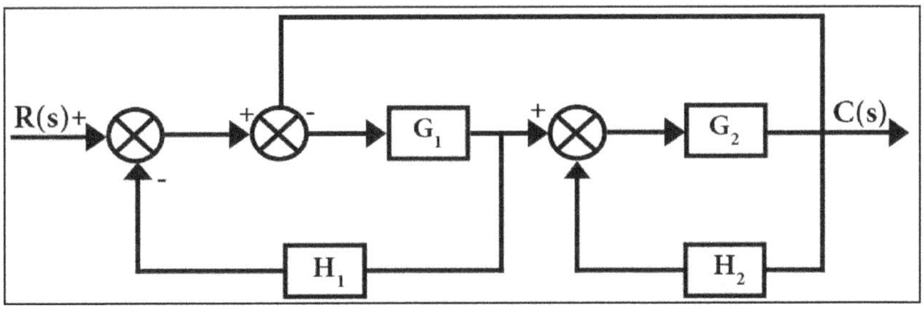

Solution:

Step 1: Eliminate the closed loop formed by G_2 and H_2, then the circuit diagram becomes,

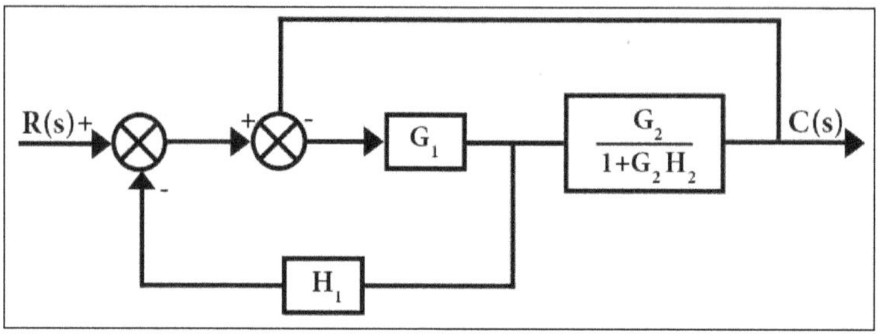

Step 2: Shift the takeoff point H_1 before the block $\frac{G_2}{1+G_2H_2}$ to a point after the block,

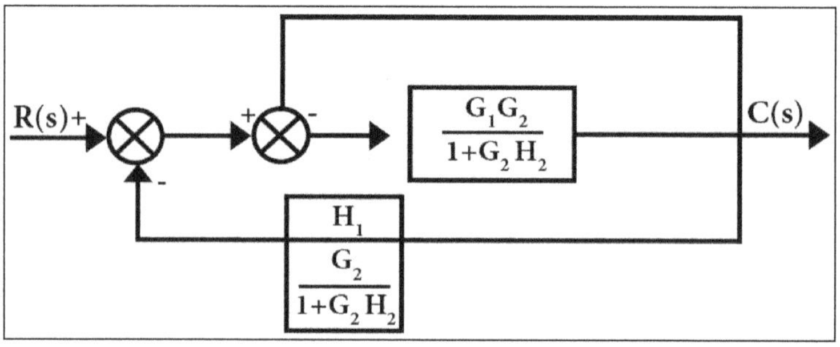

Step 3: Combine the cascade blocks G_1 and $\frac{G_2}{1+G_2H_2}$, then,

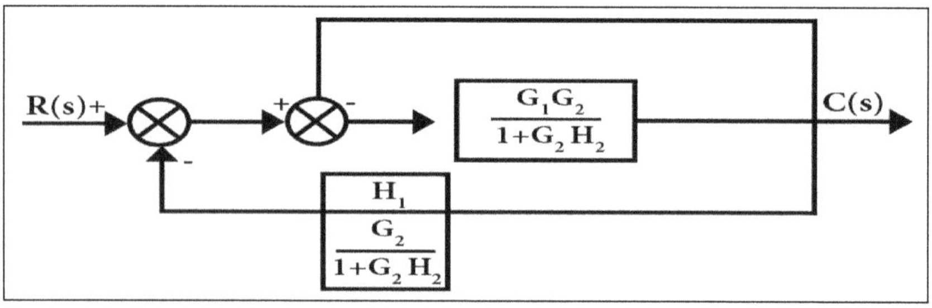

Step 4: Eliminate the Closed loop formed by $\frac{G_1G_2}{1+G_2H_2}$, then,

$$\frac{\frac{G_1G_2}{1+G_2H_2}}{1+\frac{G_1G_2}{1+G_2H_2}}=\frac{G_1G_2}{1+G_2H_2+G_1G_2}$$

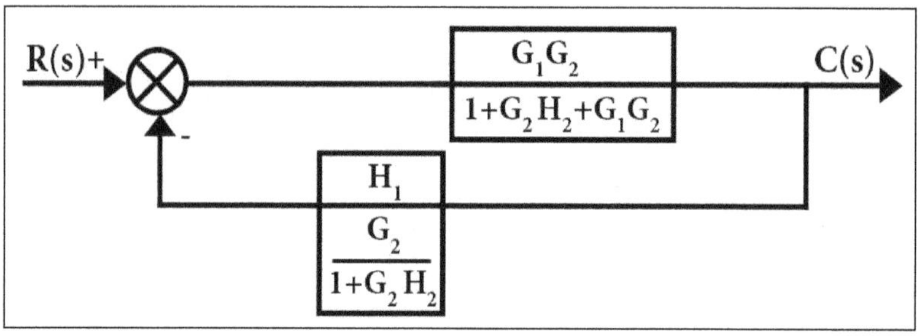

Step 5: Eliminate the closed loop formed by $\dfrac{G_1 G_2}{1+G_2 H_2+G_1 G_2}$ and $\dfrac{H_1+G_2 H_1 H_2}{G_2}$

$$\frac{C(s)}{R(s)}=\frac{\dfrac{G_1 G_2}{1+G_2 H_2+G_1 G_2}}{\dfrac{G_1 G_2}{1+G_2 H_2+G_1 G_2}\times\dfrac{H_1+G_2 H_1 H_2}{G_2}}=\frac{G_1 G_2}{1+G_2 H_2+G_1 G_2+G_1 H_1+G_1 G_2 H_1 H_2}.$$

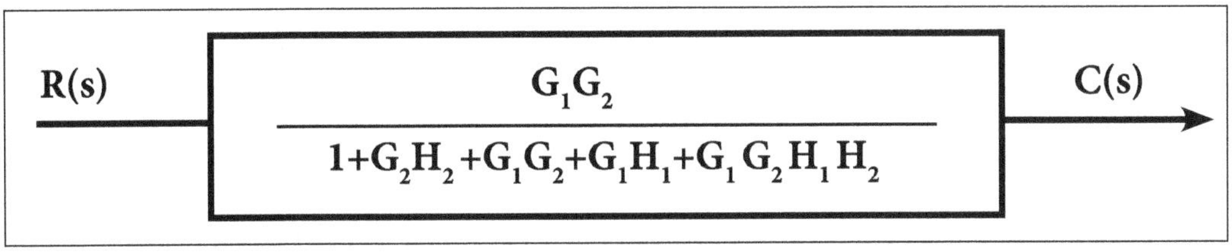

1.8 Representation by Signal Flow Graph

A signal flow graph is the diagram that represents a set of simultaneous linear algebraic equations. By taking Laplace transform, the time domain differential equations can be transferred to a set of algebraic equation in s-domain.

The signal-flow graph consists of a network in which nodes are connected by directed branches. It depicts the flow of signals from one point of a system to another and gives the relationships among the signals.

Properties of Signal Flow Graph:

- Signal flow applies only to the linear systems.
- The equations based on which a signal flow graph is drawn must be algebraic equations in the form of effects as a function of causes.
- Signals travel along the direction described by the arrows of the branches.

Procedure for determining Transfer Function using signal flow graph:

- Determine the number of forward paths and determine the forward path gain correspondingly.
- From the given graph, find the possible individual loop and corresponding gains for them. Similarly find the number of two, three etc., non-touching loops and its gain.
- Calculate the value of Δ and Δ_K. Substitute all the gains in the Mason's gain formula to determine the transfer function.

1.8.1 Reduction using Mason's Gain Formula

Mason's Gain Formula

$$T = \frac{1}{\Delta} \sum_{K} P_K \Delta_K$$

Where,

T = T (s) = Transfer function of the system.

K = Number of forward paths in the signal flow graph.

P_K = Forward path gain of K_{th} forward path.

Δ = 1 - [Sum of individual loop gains] + [Sum of gain product of all possible combination of two non-touching loops] - [Sum of gain product of all possible combination of three non-touching loops] +...

ΔK = Δ for that part of the graph which is not touching K_{th} forward path.

Comparison between Block Diagram and Signal Flow Graph Methods

S. No.	Block Diagram	Signal Flow Graph
(1)	Each element is represented by a block.	Each variable is represented by a separate node.
(2)	Transfer function of the element is given inside the block.	Transfer function is shown along the branches connecting the nodes.
(3)	Summing points and take off points are separate.	Any node can have number of incoming and outgoing branches.
(4)	Feedback path is present.	Feedback loops are considered for the analysis.
(5)	Self-loop concept is absent.	Self-loop can exist.
(6)	Applicable to known invariant systems.	Applicable to linear time invariant systems.

Problems

1. Let us derive the overall transfer function of the following block diagram using signal flow graph method.

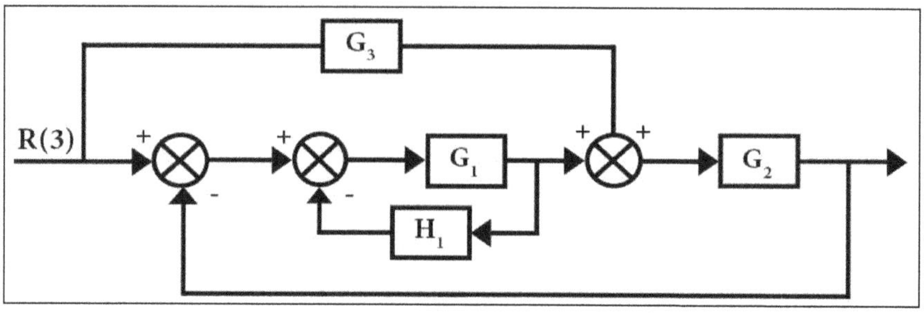

Solution:

$$\frac{C(s)}{R(s)}=\frac{G_1G_2+G_3G_2}{1+G_1G_2+G_3G_2+G_1H_1}$$

2. Let us calculate the transfer function of the transistor's hybrid model below using signal flow graph.

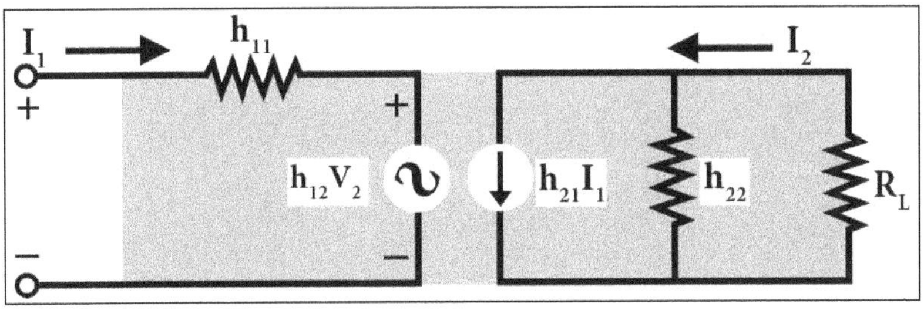

Solution:

We know that the hybrid parameter equation for a transistor is given by,

$$V_1 = h_{11}\ I_1 + h_{12}\ V_2$$

$$I_2 = h_{21}\ I_1 + h_{22}\ V_2$$

Output voltage, $V_2 = I_2$ RL

The above model can be represented in the signal flow graph as,

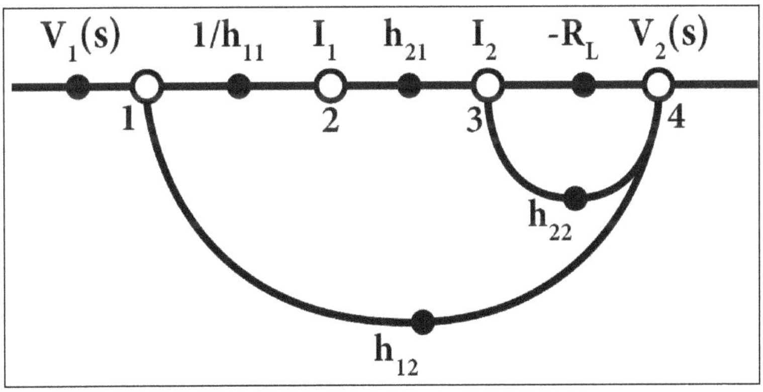

From the signal flow graph:

There is only one forward path,

$$P_1 = \frac{-h_{21}R_L}{h_{11}}$$

There are two loops such as:

Loop formed by node 1,

$$L_1 = \frac{h_{12} h_{21} R_L}{h_{11}}$$

Loop formed by node 3,

$$L_2 = h_{22} R_L$$

For all the loops touching the forward path, path factor $\Delta_{1=1}$.

$$\Delta = 1 - (L_1 + L_2) = 1 - h_{22} R_L - \frac{h_{12} h_{21} R_L}{h_{11}}$$

By using Mason's gain formula, the transfer function of transistor hybrid model is given by,

$$T.F = \frac{P_1 \Delta_1}{\Delta} = \frac{-\frac{h_{21} R_L}{h_{11}}}{1 - h_{22} R_L - \frac{h_{12} h_{21} R_L}{h_{11}}} = \frac{-h_{21} R_L}{h_{11} - h_{11} h_{22} R_L - h_{12} h_{21} R_L}$$

3. Let us calculate the C(s)/R(s) for the signal flow graph shown below.

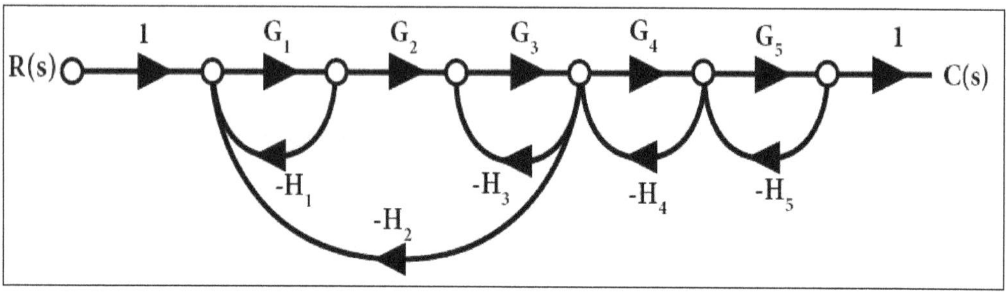

Solution:

Step 1: The number of forward path, k = 1

Forward path gain, $P_1 = G_1 G_2 G_3 G_4 G_5$

Step 2: Number of individual loops,

$$L_1 = - G_1 H_1$$

$$L_2 = - G_1 G_2 G_3 H_2$$

$$L_3 = - G_3 H_3$$

$L_4 = -G_4 H_4$

$L_5 = -G_5 H_5$

Step 3: Number of two non-touching loops at a time,

$L_1 L_3 = G_1 H_1 G_3 H_3$

$L_1 L_4 = G_1 H_1 G_4 H_4$

$L_1 L_5 = G_1 H_1 G_5 H_5$

$L_2 L_4 = G_1 G_2 G_3 H_2 G_4 H_4$

$L_2 L_5 = G_1 G_2 G_3 H_2 G_5 H_5$

$L_3 L_5 = G_3 H_3 G_5 H_5$

Step 4: There is only one, three non-touching loops at a time,

$L_1 L_3 L_5 = -G_1 H_1 G_3 H_3 G_5 H_5$

Step 5: By using mason's gain formula, the transfer function becomes,

$$\frac{C(s)}{R(s)} = \frac{\sum P_k \Delta_K}{\Delta} = \frac{P_1 \Delta_1}{1-(L_1+L_2+L_3+L_4+L_5)+(L_1L_3+L_1L_4+L_1L_5+L_2L_4+L_2L_5+L_3L_5)-(L_1L_3L_5)}$$

4. Let us obtain the transfer function of the below figure using Mason's gain formula.

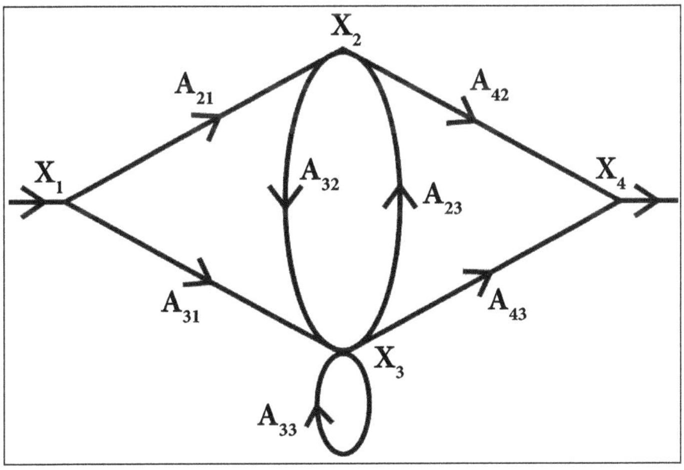

Solution:

Formula to be used:

Mason's gain formula,

$$T=\frac{1}{\Delta}\sum_{K} P_K \Delta_K$$

$$\frac{C(s)}{R(s)}=\frac{\sum_{K} P_K \Delta_K}{\Delta}=\frac{P_1 \Delta_1+P_2 \Delta_2+P_3 \Delta_3+P_4 \Delta_4}{1-[L_1+L_2]}$$

$$\frac{C(s)}{R(s)}=\frac{A_{21} A_{42}(1-A_{33})+A_{31} A_{43}+A_{21} A_{32} A_{43}+A_{31} A_{23} A_{42}}{1-(A_{23} A_{32}+A_{33})}$$

5. Let us construct the equivalent signal flow graph and C/R using Mason's gain formula for the below block diagram.

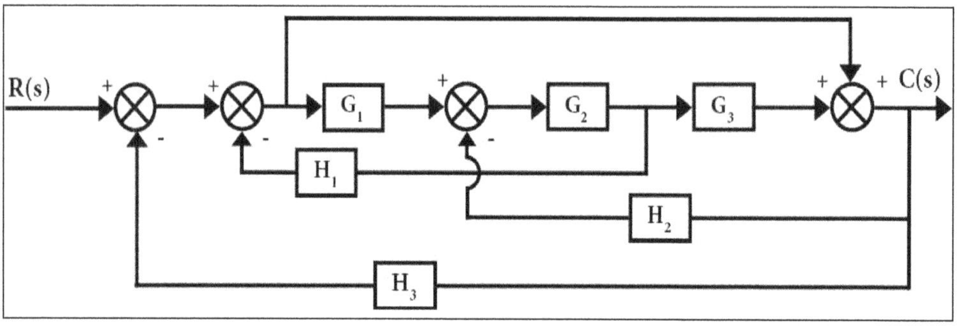

Solution:

Formula to be used:

Mason's gain formula,

$$T=\frac{1}{\Delta}\sum_{K} P_K \Delta_K$$

$$\frac{C(s)}{R(s)}=\frac{\sum_{K} P_K \Delta_K}{\Delta}=\frac{P_1 \Delta_1+P_2 \Delta_2}{1-[L_1+L_2+L_3+L_4+L_5]}$$

$$\frac{C(s)}{R(s)}=\frac{1+G_1 G_2 G_3}{1+G_1 G_2 G_3 H_3+H_3+G_1 G_2 H_1+G_2 G_3 H_2-G_2 H_1 H_2}.$$

2

Time Response Analysis

2.1 Standard Test Signals

The standard test signals are:

- Step signal.
- Ramp signal.
- Parabolic signal.
- Impulse signal.

Step Signal

Its value changes from zero to A at $t = 0$ and remains constant at A for $t \geq 0$. It resembles an actual steady input to a system. The special case of step signal is unit step in which A is unity.

The mathematical representation of the step signal is given by,

$$r(t) = u(t)$$

Where,

$$u(t) = 1;\ t > 0$$

$$u(t) = 0;\ t < 0$$

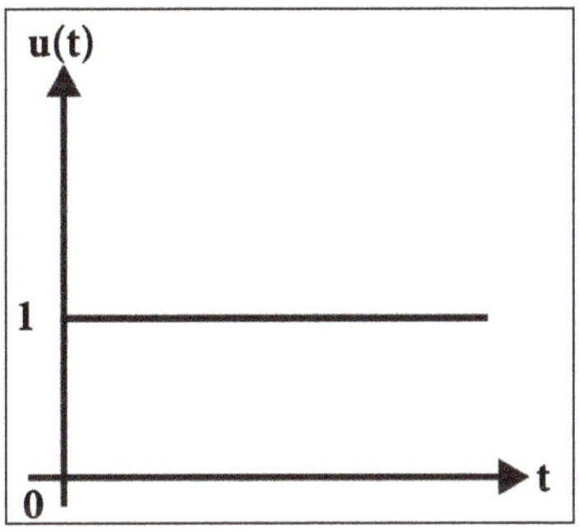

Step Signal.

Ramp Signal

Its value increases linearly with time from an initial value of zero at t = 0. It resembles a constant velocity input to the system. The special case of ramp signal is unit ramp signal in which the value of A is unity.

$$r(t) = tu\ (t)$$

Where,

$$tu\ (t) = t \text{ for } t \geq 0$$

$$= 0 \text{ else where}$$

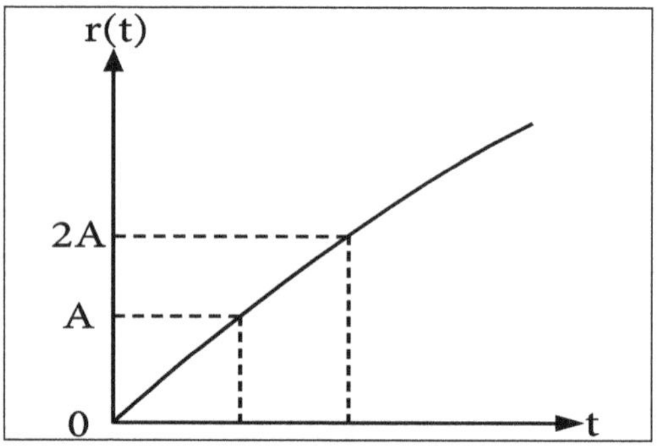

Ramp Signal.

Parabolic Signal

Here the instantaneous value varies as the square of the time from an initial value of zero at t = 0. Plot of the signal with respect to time resembles a parabola. It resembles a constant acceleration input to the system.

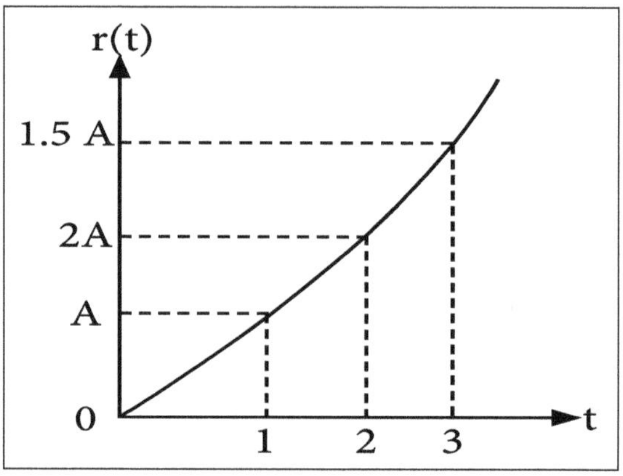

Parabolic Signal.

The special case of parabolic signal is unit parabolic signal in which A is unity. The mathematical representation of the parabolic signal is given as,

$$r(t)=\frac{At^2}{2}; t \geq 0$$

$$= 0 \quad ; t < 0.$$

Impulse Signal

It is available for very short duration. Ideal impulse signal is a unit impulse signal which is defined as a signal having zero values at all times except at t = 0.

At t = 0, the magnitude becomes infinite. It is denoted by δ (t) and mathematically it is expressed as,

$$\delta (t) = 0 \text{ for } t \neq 0$$

And,

$$\underset{t_1 \to 0}{Lt} \int_{-t_1}^{+t_1} \delta(t)dt = 1$$

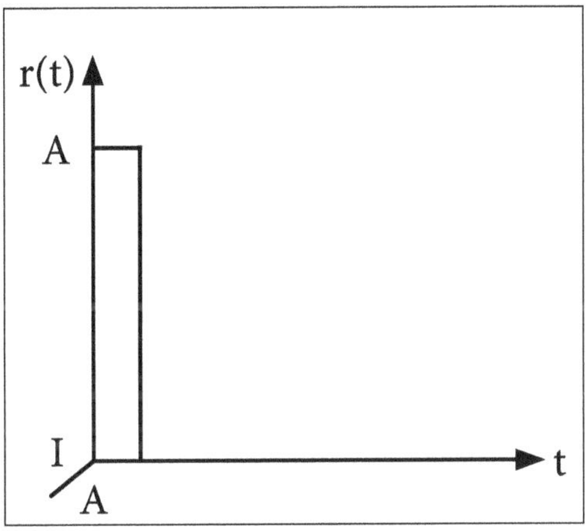

Impulse Signal.

2.1.1 Time Response of First and Second Order Systems

Unit step response of First and second order systems:

- Type (N): It is the number of poles which lies at origin.
- Order: It is the highest power of 's' in the denominator of transfer function.

First Order System

$$Ys /R(s) = K/ (1 + K + sT) = K/ (1 + sT)$$

Step Response of First Order system:

Evolution of the transient response is determined by the pole of the transfer function at s = −1/t where t is the time constant. Also, the step response can be found.

Type of the system	Order of the system	Signal waveform
Impulse response	$K/(1+sT)$	Exponential
Step response	$(K/S)-(K/(S+(1/T)))$	Step, exponential
Ramp response	$(K/S^2)-(KT/S)-(KT/(S+1/T))$	Ramp, step, exponential

LTI second-order system is expressed as,

$$G(s)=\frac{C(s)}{R(s)}=-\frac{\omega_n^2}{s^2+2\zeta\omega_n s+\omega_n^2}$$

$$(s^2+2\zeta\omega_n s+\omega_n^2)C(s)=\omega_n^2 R(s)$$

$$c(t)+2\zeta\omega_n c(t)+\omega_n^2 c(t)=\omega_n^2 r(t)$$

Expressions for the rise time, peak time, peak overshoot and settling time for the second order system.

Rise time, t_r

Put,

$$y(t)=1 \text{ at } t=t_r \text{ ,} \Rightarrow \sin(\omega s t_r+\theta)=0=\sin\pi, \Rightarrow t_r=\frac{\pi-\theta}{\omega_d};\theta=\cos-1\zeta.$$

Peak time, t_p

Put,

$$\frac{dy}{dt}=0 \text{ and solve for } t=t_p:0=\frac{\sigma\omega_n}{\omega_d}e^{-st}\sin(\omega_d t+\theta)-\omega_n e^{-st}\cos(\omega_d t+\theta)$$

$$\Rightarrow \tan(\omega_d t_p+\theta)=\frac{\omega_d}{\sigma}=\frac{\omega_n\sqrt{1-\zeta^2}}{\zeta\omega_n}=\frac{\sqrt{1-\zeta^2}}{\zeta}\tan\theta,$$

$$\Rightarrow \omega_d t_p=k\,\pi\text{ , } k=0,\ 1,\ 2,\ldots.$$

Peak overshoot occurs at k = 1 $\Rightarrow t_p = \pi / \omega_d = \pi / \omega_n \sqrt{1-\zeta^2}$.

Settling time, t_s

For 2% tolerance band, $\frac{\omega_n}{\omega_d} e^{-st} = 0.02 \Rightarrow t_s \cong \frac{4}{\sigma} = 4T$.

Steady-state error, e_{ss}

It is found that steady-state error for step input is zero. Let us now consider ramp input, r (t) = t u (t).

Then,

$$e_{ss} = \lim_{s\to 0} s\{R(s) - Y(s)\} = \lim_{s\to 0} s\left\{\frac{1}{s^2} - \frac{1}{s^2} - \frac{\omega_n^2}{s^2 + 2\zeta\omega_n s + \omega_n^2}\right\}$$

$$e_{ss} = \lim_{s\to 0} \frac{1}{s}\left\{1 - \frac{\omega_n^2}{s^2 + 2\zeta\omega_n s + \omega_n^2}\right\} = \lim_{s\to 0} \frac{1}{s}\left\{\frac{s^2 + 2\zeta\omega_n s + \omega_n^2 - \omega_n^2}{s^2 + 2\zeta\omega_n s + \omega_n^2}\right\} = \frac{2\zeta\omega_n}{\omega_n^2} = \frac{2\zeta}{\omega_n}$$

Therefore, the steady-state error due to ramp input is $\frac{2\zeta}{\omega_n}$.

Time response of second order system subjected to unit impulse input:

We know that, T.F of 2nd order system is given by,

$$\frac{C(s)}{R(s)} = \frac{\omega_n^2}{s^2 + 2\zeta\omega_n s + \omega_n^2}$$

Given input r (t) = δ (t); R (s) = 1

Substitute in the above equation,

$$C(s) = R(s)\frac{\omega_n^2}{s^2 + 2\xi\omega_n s + \omega_n^2}$$

$$C(s) = \frac{\omega_n^2}{s^2 + 2\xi\omega_n s + \omega_n^2}$$

$$C(s) = \frac{\omega_n^2}{(s + \xi\omega_n)^2 - (\xi\omega_n)^2 + \omega_n^2}$$

$$C(s) = \frac{\omega_n^2}{(s + \xi\omega_n)^2 + \omega_n^2(1 - \xi^2)}$$

$$C(s)=\frac{\omega_n^2}{(s+\xi\omega_n)^2+\sqrt{\omega_n^2(1-\xi^2)}^2}$$

$$C(s)=\frac{\omega_n^2}{(s+\xi\omega_n)^2+\left(\omega_n\sqrt{1-\xi^2}\right)^2}$$

$$C(s)=\frac{w_n^2}{w_n\sqrt{1-\xi^2}}\frac{w_n\sqrt{1-\xi^2}}{(s+\xi w_n)^2+\left(w_n\sqrt{1-\xi^2}\right)^2}$$

$$C(s)=\frac{w_n}{\sqrt{1-\xi^2}}\frac{w_n\sqrt{1-\xi^2}}{(s+\xi w_n)^2+\left(w_n\sqrt{1-\xi^2}\right)^2}$$

Take Laplace inverse transform,

$$c(t)=\frac{w_n}{\sqrt{1-\xi^2}}e^{-\xi w_n t}\sin\left(w_n\sqrt{1-\xi^2}\right)t$$

Time response of second order system subjected to unit step input:

We know that, T.F of 2nd order system is given by,

$$\frac{C(s)}{R(s)}=\frac{\omega_n^2}{s^2+2\xi\omega_n s+\omega_n^2}$$

Given input r(t) = 1; R(s) = 1/s substitute in the above equation,

$$C(s)=\frac{1}{s}-\left[\frac{s+\xi w_n}{(s+\xi w_n)^2+w_s^2(1-\xi^2)}+\frac{\xi w_n}{(s+\xi w_n)^2+w_s^2(1-\xi^2)}\right]$$

$$C(s)=\frac{1}{s}-\left[\frac{s+\xi w_n}{(s+\xi w_n)^2+\sqrt{w_n^2(1-\xi^2)}^2}+\frac{\xi w_n}{(s+\xi w_n)^2+\sqrt{w_n^2(1-\xi^2)}^2}\right]$$

$$C(s)=\frac{1}{s}-\left[\frac{s+\xi w_n}{(s+\xi w_n)^2+\left(w_n\sqrt{1-\xi^2}\right)^2}+\frac{\xi w_n}{w_n\sqrt{1-\xi^2}}\frac{w_n\sqrt{1-\xi^2}}{(s+\xi w_n)^2+\left(w_n\sqrt{1-\xi^2}\right)^2}\right]$$

Take Laplace inverse transform, then,

$$c(t)=1-\left[e^{-\xi w_n t}\cos\left(w_n\sqrt{1-\xi^2}\right)t+\frac{\xi}{\sqrt{1-\xi^2}}e^{-\xi w_n t}\sin\left(w_n\sqrt{1-\xi^2}\right)t\right]$$

By taking LCM,

$$c(t)=1-\left[\frac{\sqrt{1-\xi^2}\,e^{-\xi w_n t}\cos\left(w_n\sqrt{1-\xi^2}\right)t+\xi e^{-\xi w_n t}\sin\left(w_n\sqrt{1-\xi^2}\right)t}{\sqrt{1-\xi^2}}\right]$$

By taking $\frac{e^{-\xi w_n t}}{\sqrt{1-\xi^2}}$ commonly out,

$$c(t)=1-\frac{e^{-\xi w_n t}}{\sqrt{1-\xi^2}}\left[\sqrt{1-\xi^2}\cos\left(w_n\sqrt{1-\xi^2}\right)t+\xi\sin\left(w_n\sqrt{1-\xi^2}\right)t\right]$$

Substitute in the above equation, $\sin\varphi=\sqrt{1-\xi^2},\cos\varphi=\xi$

$$c(t)=1-\frac{e^{-\xi w_n t}}{\sqrt{1-\xi^2}}\left[\sin\varphi\cos\left(w_n\sqrt{1-\xi^2}\right)t+\cos\varphi\sin\left(w_n\sqrt{1-\xi^2}\right)t\right]$$

[By using the formula sin (A + B) = sin A cos B + cos A sin B, then]

$$c(t)=1-\frac{e^{-\xi w_n t}}{\sqrt{1-\xi^2}}\sin\left(w_n\sqrt{1-\xi^2}\,t+\varphi\right)$$

Where,

$$\tan\varphi=\frac{\sin\varphi}{\cos\varphi}=\frac{\sqrt{1-\xi^2}}{\xi},c(t)=1-\frac{e^{-\xi w_n t}}{\sqrt{1-\xi^2}}\sin\left(w_n\sqrt{1-\xi^2}\,t+\varphi\right)$$

Where $\tan\varphi=\frac{\sin\varphi}{\cos\varphi}=\frac{\sqrt{1-\xi^2}}{\xi}$

Or,

$$c(t)=1-\frac{e^{-\xi w_n t}}{\sqrt{1-\xi^2}}\sin(w_d t+\varphi);$$

Where,

$$w_d = w_n\sqrt{1-\xi^2};\ \varphi = \tan^{-1}\left(\frac{\sqrt{1-\xi^2}}{\xi}\right)$$

$$C(s)=R(s)\frac{w_n^2}{s^2+2\xi w_n s+w_n^2}$$

Substitute R(s) = 1/s in the above equation, then,

$$C(s)=\frac{1}{s}*\frac{w_n^2}{s^2+2\xi w_n s+w_n^2}$$

By applying partial fraction,

$$C(s)=\frac{A}{s}+\frac{Bs+C}{s^2+2\xi w_n s+w_n^2}$$

$$w_n^2 = A\left(s^2+2\xi w_n s+w_n^2\right)+Bs^2+Cs$$

Equate constant on both sides,

$$w_n^2 = Aw_n^2$$

$$\Rightarrow A = 1$$

Equate 's^2' term on both sides,

$$0 = A + B$$

$$\Rightarrow B = -1$$

Equate 's' term on both the sides,

$$0 = 2\xi w_n A + C$$

$$\Rightarrow C = -2\xi wn$$

Substitute A, B & C values in the above equation,

$$C(s)=\frac{1}{s}+\frac{-s-2\xi w_n}{s^2+2\xi w_n s+w_n^2}$$

$$C(s)=\frac{1}{s}-\frac{s+2\xi w_n}{s^2+2\xi w_n s+w_n^2}$$

The above equation can be rewritten as,

$$C(s)=\frac{1}{s}-\frac{s+\xi w_n+\xi w_n}{(s+\xi w_n)^2-(\xi w_n)^2 w_n^2}$$

$$C(s)=\frac{1}{s}-\frac{s+\xi w_n+\xi w_n}{(s+\xi w_n)^2+w_n^2(1-\xi^2)}$$

Error, e(t) = r(t) - c(t)

$$e(t)=1-1+\frac{e^{-\xi w_n t}}{\sqrt{1-\xi^2}}\sin(w_d t+\varphi)$$

Steady state error,

$$e_{ss}=\lim_{t=\infty}\frac{e^{-\xi w_n t}}{\sqrt{1-\xi^2}}\sin(w_d t+\varphi)=0$$

Response of second order system for under - damped case $(0 < \zeta < 1)$ and when the input is unit step:

$$\frac{Y(s)}{R(s)}=\frac{\omega_n^2}{s^2+2\zeta\omega_n s+\omega_n^2};0<\zeta<1.$$

For under damped system, $0<\zeta<1$ and the roots of the denominator are complex conjugate. The roots of the denominator are:

$$s=-\zeta\omega_n\pm\omega_n\sqrt{\zeta^2-1}$$

Since,

$$\zeta<1,\ \zeta^2<1,$$

$$s=-\zeta\omega_n\pm\omega_n\sqrt{(-1)(1-\zeta^2)}=-\zeta\omega_n\pm j\omega_n\sqrt{1-\zeta^2}$$

The damped frequency of oscillation,

$$\omega_d=\omega_n\sqrt{1-\zeta^2}$$

$$s=-\zeta\omega_n \pm j\omega_d$$

The response in s-domain,

$$C(s)=R(s)\cdot\frac{\omega_n^2}{s^2+2\zeta\omega_n s+\omega_n^2}$$

For unit step input, r(t) = 1 and $R(s)=\frac{1}{s}$

$$\therefore C(s)=\frac{\omega_n^2}{s\left(s^2+2\zeta\omega_n s+\omega_n^2\right)} \quad ...(1)$$

By partial fraction,

$$C(s)=\frac{A}{s}+\frac{B_s+C}{s^2+2\zeta\omega_n s+\omega_n^2}$$

$$A=sC(s)/s=0=\frac{\omega_n^2}{\omega_n^2}=1$$

Similarly solving for B = – 1

$$C=-\ 2\zeta\omega_n$$

$$\therefore\ C(s)=\frac{1}{s}-\frac{(s+2\zeta\omega_n)}{s^2+2\zeta\omega_n\, s+\omega_n^2} \quad ...(2)$$

Adding and subtracting $\zeta^2\omega^2 n$ to the denominator of second term in the equation (2),

$$C(s)=\frac{1}{s}-\frac{(s+2\zeta\omega_n)}{s^2+2\zeta\omega_n\, s+\omega_n^2+\zeta^2 w_n^2-\zeta^2\omega_n^2}$$

$$C(s)=\frac{1}{s}-\frac{(s+2\zeta\omega_n)}{\left(s^2+2\zeta\omega_n\, s+\zeta^2 w_n^2\right)+\left(\omega_n^2-\zeta^2\omega_n^2\right)} C(s)=\frac{1}{s}-\frac{(s+2\zeta\omega_n)}{(s+\zeta\omega_n)^2+\omega_n^2\left(1-\zeta^2\right)}$$

$$C(s)=\frac{1}{s}-\frac{\left(s+2\zeta\omega_n^1\right)}{(s+\zeta\omega_n)^2+\omega_n^2}\left[\because\ \omega_d=\omega_n\sqrt{1-\zeta^2}\right]$$

$$C(s)=\frac{1}{s}-\frac{(s+2\zeta\omega_n)}{(s+\zeta\omega_n)^2+\omega_d^2}-\frac{\zeta\omega_n}{(s+\zeta\omega_n)^2+\omega_d^2} \quad ...(3)$$

Multiplying and dividing by ω_d in the third term of the equation (3),

$$C(s)=\frac{1}{s}-\frac{(s+2\zeta\omega_n)}{(s+\zeta\omega_n)^2+\omega_d^2}-\frac{\zeta\omega_n}{\omega_d}-\frac{\omega_d}{(s+\zeta\omega_n)^2+\omega_d^2}$$

On taking L.T,

$$C(t)=1-e^{\zeta\omega t}\cos\omega_a t-\frac{\zeta\omega_n}{\omega_d}-e^{-\zeta\omega t}\sin\omega_d t$$

$$C(t)=1-e^{-\zeta\omega t}\left[\cos\omega_a t+\frac{\zeta\omega_n}{\omega_d\sqrt{1-\zeta^2}}-\sin\omega_d t\right]$$

$$C(t)=\frac{1-e^{-\zeta\omega t}}{\sqrt{1-\zeta^2}}\left[\cos\omega_a t+\left(\sqrt{1-\zeta^2}\right)+\zeta\sin\omega_d t\right]$$

$$C(t)=\frac{1-e^{-\zeta\omega t}}{\sqrt{1-\zeta^2}}\left[(\sin\omega_d t\cdot\cos\theta+\cos\omega_d t\cdot\sin\theta)\right]$$

$$C(t)=\frac{1-e^{-\zeta\omega t}}{\sqrt{1-\zeta^2}}\left[\sin(\omega_d t+\theta)\right]$$

Note:

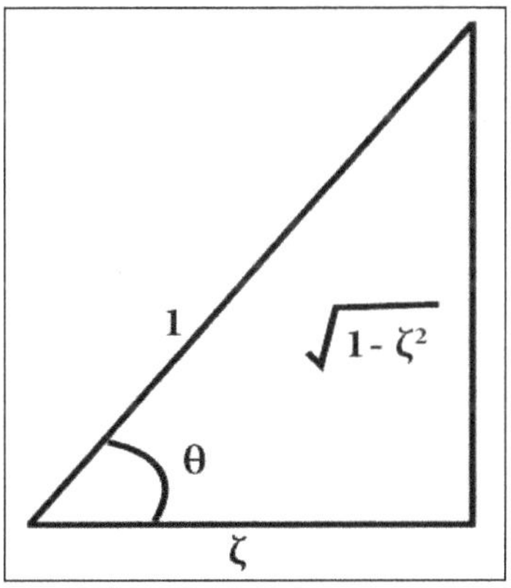

$$\sin\theta=\sqrt{1-\zeta^2}$$

$$\cos\theta=\zeta$$

$$\tan\theta=\frac{\sqrt{1-\zeta^2}}{\zeta}$$

Where,

$$\theta = \tan^{-1}\frac{\sqrt{1-\zeta^2}}{\zeta}$$

The response oscillates before settling to a final value. The oscillation depends on the value of damping ratio.

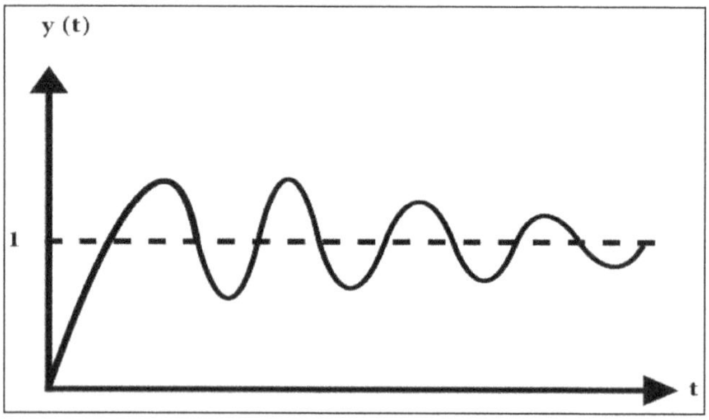

2.1.2 Time Domain Specifications

The time response of the control system consists of two parts such as the transient response and the steady state response. The transient response is defined as the part of the time response which goes from the initial state to the final state and reduces to zero as time becomes very large. The steady-state response is defined as the behavior of the system as approaches infinity after the transients have died out. Thus, the system response y (t) may be written as,

$$y(t) = yt(t) + y_{ss}(t)$$

Where,

y_t (t) denotes the transient response and y_{ss} (t) denotes the steady-state response.

The control systems are generally designed with damping less than one. i.e., Oscillatory step response. Higher order control systems usually have a pair of complex conjugate poles with damping less than unity which dominate over the other poles. In specifying the transient-response characteristics of the control system to the unit step input, we usually specify the following:

- Rise time, t_r
- Delay time, t_d
- Peak overshoot, M_P

- Peak time, t_p
- Steady-state error, e_{ss}
- Settling time, t_s

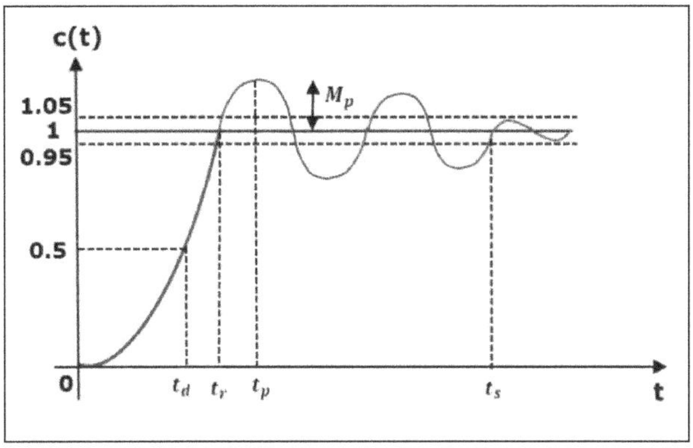

Time Domain Specifications.

Rise Time, t_r: It is the time required for the response to rise from 0 to 100% of the final value for the under damped system.

Delay Time, t_d: It is the time required for the response to reach 50% of the final value in first attempt.

Settling Time, t_s: It is the time required for the response to reach and stay within a specified tolerance band (2% or 5%) of its final value.

Peak Overshoot, M_p: It is the normalized difference between the time response peak and the steady output and is defined as,

$$\%M_p = \frac{c(t_p) - c(\infty)}{c(\infty)} \times 100\%$$

Peak Time, t_p: Peak time is the time required for the response to reach the peak time response or the peak overshoot.

Steady-State Error, e_{ss}: It indicates the error between the actual output and desired output as 't' tends to infinity.

$$e_{ss} = \lim_{t\to} \left[r(t) - c(t) \right]$$

Problems

1. Let us determine the open loop transfer function of the unit impulse response for a unit feedback control system. c (t) = $-t\,e^{-t} + 2e^{-t}$, ($t \geq 0$).

Solution:

Given:

$$c(t) = -t\,e^{-t} + 2e^{-t}, (t \geq o)$$

Formula to be used:

Transfer function of unity feedback system $\frac{C(s)}{R(s)} = \frac{G(s)}{1+G(s)}$

Open loop transfer function $G(s) = \frac{C(s)}{1-C(s)}$

$$c(t) = -t\,e^{-t} + 2e^{-t}$$

Take Laplace transform,

$$C(s) = -\frac{1}{(s+1)^2} + 2\frac{1}{s+1}$$

$$C(s) = \frac{-1+2(s+1)}{(s+1)^2}$$

$$C(s) = \frac{-1+2s+2}{(s+1)^2}$$

$$C(s) = \frac{2s+1}{(s+1)^2}$$

Given input r (t) = δ (t), R (S) = 1

We know that, Transfer function of unity feedback system,

$$\frac{C(s)}{R(s)} = \frac{G(s)}{1+G(s)}$$

$$\frac{C(s)}{1} = \frac{G(s)}{1+G(s)}$$

$$C(s)[1 + G(s)] = G(s)$$

$$G(s) - G(s)C(s) = C(S)$$

$$G(s)[1-C(s)]=C(s)$$

Open loop transfer function,

$$G(s)=\frac{C(s)}{1-C(s)}$$

$$G(s)=\frac{\frac{2s+1}{(s+1)^2}}{1-\frac{2s+1}{(s+1)^2}}$$

$$G(s)=\frac{2s+1}{(s+1)^2-[2s+1]}$$

$$G(s)=\frac{2s+1}{s^2+2s+1-2s-1}$$

$$G(s)=\frac{2s+1}{s^2}$$

2. Let us obtain an expression for unit step response of the system $G(s)=\frac{2}{s(s+3)}$.

Solution:

Given:

$$G(s)=\frac{2}{s(s+3)}$$

Formula to be used:

Transfer function of unity feedback system,

$$\frac{C(s)}{R(s)}=\frac{G}{1+G(s)}$$

The closed loop transfer function is given by,

$$\frac{C(s)}{R(s)}=\frac{G}{1+G(s)}=\frac{\frac{2}{s(s+3)}}{1+\frac{2}{s(s+3)}}=\frac{2}{s^2+3s+2}$$

Given input r (t) = 1 ⇒ R (s) = 1/s, then,

$$C(s)=R(s)\frac{2}{s^2+3s+2}=\frac{2}{s(s^2+3s+2)}=\frac{A}{s}+\frac{B}{s+1}+\frac{C}{s+2}$$

$$2=A(s+1)(s+2)+Bs(s+2)+Cs(s+1)$$

Equating constants on bot h sides, 2 = 2A ⇒ A = 1

Substitute s = -1, then 2 = B (-1) (1) ⇒ B = -2

Substitute s = -2, then 2 = C (-2) (-1) ⇒ C = 1

$$C(s)=\frac{A}{s}+\frac{B}{s+1}+\frac{C}{s+2}=\frac{1}{s}-\frac{2}{s+1}+\frac{1}{s+2}$$

Take inverse Laplace transform on both the sides.

The expression for unit step response becomes, $c(t)=1-2e^{-t}\ e^{-2t}$.

3. The unity feedback system is characterized by an open loop transfer function $G(s)=\frac{K}{s(s+10)}$. Let us determine the gain K, so that the system will have a damping ratio of 0.5. For this value of K, also calculate the settling time, peak over shoot gain K, time to peak overshoot for a unit step input.

Solution:

Given:

$$G(s)=\frac{K}{s(s+10)}$$

Damping ratio, $\xi=0.5$

Formula to be used:

$$\frac{C(s)}{R(s)}=\frac{G(s)}{1+G(s)H(s)}$$

$$\frac{C(s)}{R(s)}=\frac{w_n^2}{s^2+2\xi w_n s+w_n^2}$$

Peak time,

$$t_p = \frac{\pi}{w_n\sqrt{1-\xi^2}}$$

Maximum peak overshoot,

$$M_p\% = e^{-\xi\pi\sqrt{1-\xi^2}} * 100\%$$

Given damping ratio, $\xi = 0.5$

Closed loop transfer function of a 2nd order system,

$$\frac{C(s)}{R(s)} = \frac{G(s)}{1+G(s)H(s)} = \frac{\frac{K}{s(s+10)}}{1+\frac{K}{s(s+10)}\times 1} = \frac{K}{s(s+10)+K} = \frac{K}{s^2+10s+K}$$

By comparing with $\frac{C(s)}{R(s)} = \frac{w_n^2}{s^2+2\xi w_n s + w_n^2}$

$$w_n^2 = K \Rightarrow w_n = \sqrt{K} \text{ rad/sec}$$

$$2\xi w_n = 10 w_n = \frac{10}{2\xi_n} = \frac{10}{2\times 0.5} = 10$$

Substitute w_n in the above equation, we get,

$$K = w_n^2 = 10^2 = 100$$

(i) Peak time:

$$t_p = \frac{\pi}{w_n\sqrt{1-\xi^2}} = \frac{3.14}{10\times\sqrt{0.75}} = \frac{3.14}{8.66} = 0.362 \text{ sec}$$

(ii) Maximum peak overshoot:

$$M_p\% = e^{-\xi\pi\sqrt{1-\xi^2}} * 100\%$$

$$= e^{-0.5\pi/0.866} \times 100\% \Rightarrow 16.3\%$$

(iii) Setting time t_s for $2\% = \frac{4}{\xi w_n} = \frac{4}{0.5\times 10} = \frac{4}{5} = 0.8 \text{ sec.}$

4. The open loop transfer function of a unity feedback system is given by $G(s) = \frac{K}{s(sT+1)}$

, where K and T are positive constants. Let us analyze by what factor the amplifier gain K should be reduced, so that the peak overshoot of unit step response of the system is reduced from 75% to 25%.

Solution:

Given:

$$G(s)=\frac{K}{s(sT+1)}$$

Formula to be used:

$$\frac{C(s)}{R(s)}=\frac{G(s)}{1+G(s)H(s)}$$

$$\xi=\frac{1}{2\sqrt{KT}}$$

$$G(s)=\frac{K}{s(sT+1)} \text{ and } H(s)=1$$

The closed loop transfer function becomes,

$$\frac{C(s)}{R(s)}=\frac{G(s)}{1+G(s)H(s)}=\frac{\frac{K}{s(sT+1)}}{1+\frac{K}{s(sT+1)}\times 1}=\frac{K}{sT^2+s+K}$$

The above equation can be rewritten as,

$$\frac{C(s)}{R(s)}=\frac{K}{T^2\left(s^2+\frac{s}{T^2}+\frac{K}{T^2}\right)}=\frac{\frac{K}{T}}{s^2+\frac{s}{T}+\frac{K}{T}}$$

By comparing with 2nd order system equation $s^2+2\xi w_n s+w_n^2$.

Then,

$$w_n^2=\frac{K}{T}\Rightarrow w_n=\sqrt{\frac{K}{T}}\text{ rad/sec}$$

$$2\xi w_n=\frac{1}{T}=\xi=\frac{1}{2w_nT}=\frac{1}{2\sqrt{\frac{K}{T}}T}=\frac{1}{2\sqrt{KT}}$$

Let,

$$\xi = \xi_1 \text{ for Mp} = 75\% \text{ or } 0.75$$

$$\therefore\ 0.75 = e^{-\frac{\xi_1 \pi}{\sqrt{1-\xi_1^2}}}$$

Taking log on both sides,

$$\text{In}(0.75) = \text{In}\, e^{-\frac{\xi_1 \pi}{\sqrt{1-\xi_1^2}}} \Rightarrow -0.2876\sqrt{1-\xi_1^2} = -3.14\xi_1 \Rightarrow \sqrt{1-\xi_1^2} = 10.91\,\xi_1$$

Taking square on both sides,

$$1-\xi_1^2 = 119.20$$

$$\xi_1^2 \Rightarrow \xi_1 = 0.09121$$

Let,

$$\xi = \xi_2 \text{ for } M_p = 25\% \text{ or } 0.25$$

$$\therefore\ 0.25 = e^{-\frac{\xi_2 \pi}{\sqrt{1-\xi_2^2}}}$$

Taking log on both sides,

$$\text{In}(0.25) = \text{In}\ e^{-\frac{\xi_2 \pi}{\sqrt{1-\xi_2^2}}} \Rightarrow -1.386\sqrt{1-\xi_2^2} = -3.14\xi_2 \Rightarrow \sqrt{1-\xi_2^2} = 2.265\,\xi_2$$

Taking square on both sides,

$$1-\xi_2^2 = 5.13\,\xi_2^2 \Rightarrow 6.13\,\xi_2^2 = 1 \Rightarrow \xi_2 = \sqrt{\frac{1}{6.13}} = 0.40389$$

We know that,

$$\xi = \frac{1}{2\sqrt{KT}}$$

$$\frac{\xi_1}{\xi_2} = \frac{\frac{1}{2\sqrt{K_1T}}}{\frac{1}{2\sqrt{K,T}}} = \frac{0.09121}{0.40389} = 0.2258$$

$$\frac{K_2}{K_1}=0.05099$$

$$K_2=0.05099\ K_1 \text{ or } K_2=\frac{1}{19.4}K_1$$

5. The closed loop transfer function of a second order system which is given by $T(s)=\frac{100}{s^2+10s+100}$. Let us determine the damping ratio, natural frequency of oscillations, rise time, settling time and peak overshoot for the closed loop transfer function.

Solution:

Given:

$$T(s)=\frac{100}{s^2+10s+100}$$

Formula to be used:

$$\frac{C(s)}{R(s)}=\frac{w_n^2}{s^2+2\xi w_n s+w_n^2}$$

Rise time $t_r=\frac{\pi-\phi}{w_n\sqrt{1-\xi^2}}$

Peak time $t_p=\frac{\pi}{w_n\sqrt{1-\xi^2}}$

Maximum peak overshoot $M_p\%=e^{-\xi\pi/\sqrt{1-\xi^2}}*100\%$

Settling time t_s for $2\%=\frac{4}{\xi w_n}$

Closed loop transfer function of a 2nd order system,

$$\frac{C(s)}{R(s)}=\frac{100}{s^2+10s+100}$$

By comparing with $\frac{C(s)}{R(s)}=\frac{w_n^2}{s^2+2\xi w_n s+w_n^2}$

We get,

$$w_n^2=100\Rightarrow w_n=10\,\text{rad/sec}$$

$$2\xi w_n = 10 \Rightarrow \xi = \frac{10}{2w_n} = 0.5$$

Rise time $$t_r = \frac{\pi - \phi}{w_n\sqrt{1-\xi^2}} = \frac{\pi - \tan^{-1}\left(\frac{\sqrt{1-\xi^2}}{\xi}\right)}{w_n\sqrt{1-\xi^2}} = \frac{\pi - 1.047}{10 x \sqrt{0.75}} = 0.241 sec.$$

Peak time $$t_p = \frac{\pi}{w_n\sqrt{1-\xi^2}} = \frac{3.14}{10 x \sqrt{0.75}} = \frac{3.14}{8.66} = 0.362 sec$$

Maximum peak overshoot,

$$M_p \% = e^{-\xi\pi/\sqrt{1-\xi^2}} * 100\%$$

$$= e^{-0.5\pi/0.866} * 100\% \Rightarrow 16.3\%$$

Settling time t_s for 2% $$= \frac{4}{\xi w_n} = \frac{4}{0.5 \times 10} = \frac{4}{5} = 0.8 sec$$

6. Let us analyze by what factor the amplifier gain K can be reduced, so that the peak overshoot of unit step response of the system is reduced from 85% to 35%. It is given that the open loop transfer function of a unity feedback system is G(s) = K/s (sT + 1) where K and T are positive constants.

Solution:

Given:

$$G(s) = K / s(sT + 1)$$

Formula to be used:

$$\frac{C(s)}{R(s)} = \frac{G(s)}{1 + G(s)H(s)}$$

$$\xi = \frac{1}{2\sqrt{KT}}$$

$$G(s) = \frac{K}{s(sT+1)} \text{ and } H(s) = 1$$

The closed loop transfer function becomes,

$$\frac{C(s)}{R(s)}=\frac{G(s)}{1+G(s)H(s)}=\frac{\frac{K}{s(sT+1)}}{1+\frac{K}{s(sT+1)}*1}=\frac{K}{sT^2+s+K}$$

The above equation can be rewritten as,

$$\frac{C(s)}{R(s)}=\frac{K}{T^2\left(s^2+\frac{s}{T^2}+\frac{K}{T^2}\right)}=\frac{\frac{K}{T}}{s^2\frac{s}{T}+\frac{K}{T}}$$

By comparing with 2nd order system equation $s^2+2\xi w_n s+w_n^2$.

Then,

$$w_n^2=\frac{K}{T}\Rightarrow w_n=\sqrt{\frac{K}{T}}\text{ rad/sec}$$

$$2\xi w_n=\frac{1}{T}\Rightarrow\xi=\frac{1}{2w_n T}=\frac{1}{2\sqrt{\frac{K}{T}}T}=\frac{1}{2\sqrt{KT}}$$

Let,

$$\xi=\xi_1 \text{ for } M_p = 85\% \text{ or } 0.85$$

$$\therefore\ 0.85=e^{-\frac{\xi_1\pi}{\sqrt{1-\xi_1^2}}}$$

Taking log on both sides,

$$\text{In}(0.85)=\text{In } e^{-\frac{\xi_1\pi}{\sqrt{1-\xi_1^2}}}\Rightarrow -1.162\sqrt{1-\xi_1^2}=-3.14\xi_1\Rightarrow\sqrt{1-\xi_1^2}=19.33\,\xi_1$$

Taking square on both sides,

$$1-\xi_1^2=373.64\xi_1^2\Rightarrow\xi_1=0.0517$$

Let,

$$\xi=\xi_2 \text{ for } M_p = 35\ \% \text{ or } 0.35$$

$$\therefore\ 0.35=e^{-\frac{\xi_2\pi}{\sqrt{1-\xi_2^2}}}$$

Taking log on both sides,

$$\ln(0.35)=\ln e^{-\frac{\xi_2 \pi}{\sqrt{1-\xi_2^2}}} \Rightarrow -1.0498\sqrt{1-\xi_2^2}=-3.14\xi_2 \Rightarrow \sqrt{1-\xi_2^2}=2.99\,\xi_2$$

Taking square on both sides,

$$1-\xi_2^2=8.94\xi_2^2 \Rightarrow \xi_2=0.317$$

We know that,

$$\xi=\frac{1}{2\sqrt{KT}}$$

$$\frac{\xi_1}{\xi_2}=\frac{\frac{1}{2\sqrt{K_1T}}}{\frac{1}{2\sqrt{K_2T}}}=\frac{0.0517}{0.317}=0.163$$

$$\frac{K_2}{K_1}=0.026569$$

$$K_2=0.026569K_1$$

Therefore, the required factor is $0.026569K_1$.

7. An unity feedback control system is shown in the below figure. By using derivative control, the damping ratio is to be made to 0.8. Let us determine the value of T_d and compare the rise time, peak time and maximum overshoot of the system. (i) Without derivative control, (ii) With derivative control. The input to the system is unit step.

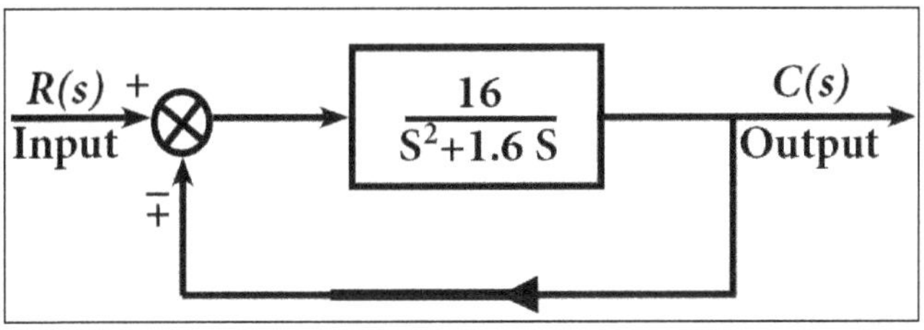

Solution:

Unit step input:

Damping ratio = 0.8

Formula to be used:

$$\frac{C(s)}{R(s)} = \frac{w_n^2}{s^2 + 2\xi w_n s + w_n^2}$$

$$\text{Rise time } t_r = \frac{\pi - \phi}{w_n \sqrt{1-\xi^2}}$$

$$\text{Peak time } t_p = \frac{\pi}{w_n \sqrt{1-\xi^2}}$$

$$\text{Maximum peak overshoot } M_p\% = e^{-\xi\pi/\sqrt{1-\xi^2}} * 100\%$$

$$\text{Settling time } t_s\ 2\% = \frac{4}{\xi w_n}$$

Closed loop transfer function of the system,

$$\frac{C(s)}{R(s)} = \frac{16}{s^2 + 1.6s + 16}$$

Without derivative control,

By comparing with $\frac{C(s)}{R(s)} = \frac{w_n^2}{s^2 + 2\xi w_n s + w_n^2}$

$$w_n^2 = 16 \Rightarrow w_n = 4 \text{ rad/sec}$$

$$2\xi w_n = 1.6 \Rightarrow \xi = \frac{1.6}{2w_n} = 0.2$$

i. Rise time $t_r = \frac{\pi - \phi}{w_n \sqrt{1-\xi^2}} = \frac{\pi - \tan^{-1}\left(\frac{\sqrt{1-\xi^2}}{\xi}\right)}{w_n \sqrt{1-\xi^2}} = 0.45 \text{ sec}$

ii. Peak time $t_p = \frac{\pi}{w_n \sqrt{1-\xi^2}} = 0.8 \text{ sec}$

iii. Maximum peak overshoot $M_p\% = e^{-\xi\pi/\sqrt{1-\xi^2}} * 100\% \Rightarrow 52.6\%$

iv. Settling time $t_s\ 2\% = \frac{4}{\xi w_n} = \frac{4}{0.2 \times 4} = 5 \text{ sec}$

With derivative, the new damping ratio is 0.8

$$\xi' = \xi + \frac{w_n T_d}{2} \Rightarrow 0.8 = 0.2 + \frac{4T_d}{2}$$

The value of $= T_d = \frac{0.8 - 0.2}{2} = 0.3$

8. Let us determine the expression for closed-loop transfer on a servo mechanisms which show the system response as c (t) = 1 + 0. $2e^{-60t}$ – $1.2e^{-10t}$ when subjected to a unit step input. Let us determine the expression for closed-loop transfer function and also obtain the un damped natural frequency and damping ratio function.

Solutions:

Given:

$$c(t) = 1 + 0.2\,e^{-60t} - 1.2e^{-10t}$$

By taking Laplace Transform on both the sides,

$$C(s) = \frac{1}{s} + \frac{0.2}{s+60} - \frac{1.2}{s+10} = \frac{(s+60)(s+10) + 0.2(s+10) - 1.2s(s+60)}{s+(s+60)(s+10)}$$

$$C(s) = \frac{600}{s(s^2+70s+600)}$$

Given input r(t) =1, R(s) $= \frac{1}{s}$

i. The closed loop Transfer function $\frac{C(s)}{R(s)} = \frac{600}{s^2+70s+600}$

ii. Undammed natural function $w_n^2 = 600 \Rightarrow w_n = \sqrt{600} = 24.49 \text{ rad/sec}$

Damping ratio $\xi = \frac{70}{2w_n} = \frac{70}{48.98} = 1.43$

2.2 Steady State Errors and Error Constants

There are Two Types of Error Coefficients:

- Static error coefficient.
- Dynamic error coefficient.

Static Error Coefficient

Ability of the system to reduce or eliminate the steady state error is static error coefficient. It is of 3 different types:

Position Error Coefficient:

Here the reference input signal is a constant the output signal is a constant in steady-state. The error constant associated with this condition is then referred as the position error constant. It is given the symbol K_p.

Velocity Error Coefficient:

Here the reference input is a ramp then the output position signal is a ramp signal in steady-state. The signal that is constant in this situation is the velocity, which is the derivative of the output position. The error constant is referred as the velocity error constant. It is given the symbol K_v.

Acceleration Error Coefficient:

Here the reference input is a parabola, and then the output position signal is also a parabola (constant curvature) in steady-state. Therefore, the signal that is constant in this situation is the acceleration, which is the second derivative of the output position. The error constant is referred as the acceleration error constant. It is given the symbol K_a.

Type	Step input	Ramp input	Parabolic input
Type 0 system	1/(1+Kp)	∞	∞
Type 1 system	0	1/Kv	∞
Type 2 system	0	0	1/Ka

Features of static error are as follows:

- Higher error coefficient increases the steady state performance of the system.
- It cannot be used to calculate the error of unstable system.
- It does not indicate the correct manner in which the error changes with time.
- It is difficult to stabilize the system.

Dynamic Error Coefficient

It is used to express the dynamic error. It provides the error signal as a function of time. It is used for determining any type of input. The dynamic error coefficient provides a simple way of estimating the error signal.

Steady State Errors

One of the important design specifications for a control system is the steady state error. The steady state output of any system should be as close to desired output as possible. If it deviates from this desired output, the performance of the system is not satisfactory under steady state conditions.

The steady state error reflects the accuracy of the system. Among many reasons for these errors, the most important ones are the type of input, the type of the system and the non-insanities present in the system. Since the actual input in a physical system is often a random signal, the steady state errors are obtained for the standard test signals namely step, ramp and parabolic signals.

Error Constants

Let us consider a feedback control system shown in the below figure:

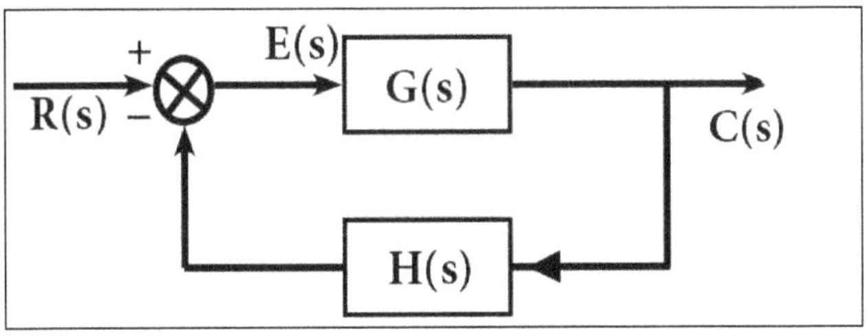

Feedback control system.

The error signal E(s) is given by,

$$E(s) = R(s) - H(s)\,C(s)$$

But $C(s) = G(s)\,E(s)$

We have,

$$E(s) = \frac{R(s)}{1+G(s)\,H(s)}$$

Applying final value theorem, we can get the steady state error e_{ss} as,

$$e_{ss} = \underset{s \to 0}{Lt}\ sE(s) = \underset{s \to 0}{Lt}\ \frac{sR(s)}{1+G(s)H(s)}$$

The above equation shows that the steady state error is a function of the input R(s) and the open loop transfer function G(s). Let us consider various standard test signals and obtain the steady state error for these inputs.

1. Unit Step or Position Input

For a unit step input $R(s) = \frac{1}{s}$. Hence from the above equation,

$$e_s = \underset{s \to 0}{Lt}\ \frac{s \cdot \frac{1}{s}}{1+G(s)H(s)}$$

$$= \frac{1}{1 + \underset{s \to 0}{\text{Lt}}\, G(s)H(s)}$$

Let us define a useful term, positional error constant K_p as,

$$K_p \triangleq \underset{s \to 0}{\text{Lt}}\, G(s)\, H(s)$$

In terms of the position error constant, e_{ss} can be written as,

$$e_{ss} = \frac{1}{1 + K_p}$$

2. Unit Ramp or Velocity Input

For unit velocity input, $R(s) = 1/s^2$ and hence,

$$e_{ss} = \underset{s \to 0}{\text{Lt}} \frac{s \cdot \frac{1}{s}}{1 + G(s)H(s)} = \underset{s \to 0}{\text{Lt}} \frac{1}{s + sG(s)H(s)}$$

$$= \frac{1}{\underset{s \to 0}{\text{Lt}}\, sG(s)H(s)}$$

Again, defining the velocity error constant K_v as,

$$K_v = \underset{s \to 0}{\text{Lt}}\, sG(s)H(s)$$

$$e_{ss} = \frac{1}{K_v}$$

Unit Parabolic or Acceleration Input

For unit acceleration input, $R(s) = 1/s^3$ and hence,

$$e_{ss} = \underset{s \to 0}{\text{Lt}} \frac{s}{s^3 + \left[1 + G(s)H(s)\right]} = \underset{s \to 0}{\text{Lt}} \frac{1}{s^2 + s^2 G(s)H(s)}$$

$$= \frac{1}{\underset{s \to 0}{\text{Lt}}\, s^2 G(s)H(s)}$$

Defining the acceleration error constant K_a as,

$$K_a = \underset{s \to 0}{Lt}\ s^2 G(s)H(s)$$

$$e_{ss} = \frac{1}{K_a}$$

For the special case of unity feedback system, H(s) = 1, the above equations are modified as,

$$K_p = \underset{s \to 0}{Lt}\ G(s)$$

$$K_v = \underset{s \to 0}{Lt}\ sG(s)$$

$$K_a = \underset{s \to 0}{Lt}\ s^2 G(s)$$

Expression to Find Steady State Error of a Closed Loop Control System

Consider the simple closed loop system using negative feedback.

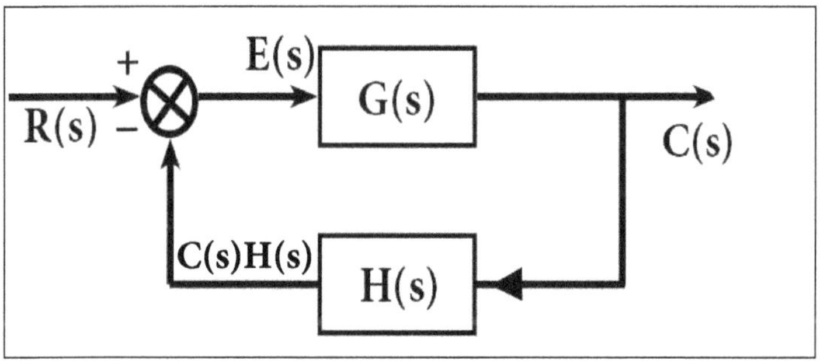

In the above figure.

G(s) – Forward path transfer function.

H(s) – Feedback transfer function.

E(s) – Error signal.

C(s) H(s) – Feedback signal

The error of the signal, E(s) = R(s) – C(s) H(s)

But,

C(s) = G(s) E(s)

Then the above equation becomes,

E(s) = R(s) – G(s) E(s) H(s)

$$E(s) + E(s)\,G(s)\,H(s) = R(s)$$

$$E(s)\,[1 + G(s)\,H(s)] = R(s)$$

Therefore, the error for non-unity feedback system is given by,

$$E(s)=\frac{R(s)}{1+G(s)\,H(s)}$$

Therefore, the error for unity feedback system is given by,

$$E(s)=\frac{R(s)}{1+G(s)}$$

By applying the final value theorem, the steady state error can be determined as,

$$e_{ss}=\lim_{s\to 0} sE(s)\,[\text{or}]\,e_{ss}=\lim_{t\to\infty} e(t)$$

The steady state error of the closed loop system is given by,

$$e_{ss}=\lim_{s\to 0} s\frac{R(s)}{1+G(s)\,H(s)}$$

Relationships Between Input, System Type, Static Error Constants and Steady State Errors

Input	Steady state error formula	Type 0		Type 1		Type 2	
		Static error constant	Error	Static error constant	Error	Static error constant	Error
Step, u(t)	$\frac{1}{1+K_p}$	K_p =constant	$\frac{1}{1+K_p}$	$K_p=\infty$	0	$K_p=\infty$	0
Ramp, $tu(t)$	$\frac{1}{K_v}$	$K_v=0$	∞	K_v =constant	$\frac{1}{K_v}$	$K_v=\infty$	0
Parabola, $1/2\,t^2\,u(t)$	$\frac{1}{K_a}$	$K_a=0$	∞	$K_a=0$	∞	K_a =constant	$\frac{1}{K_a}$

Generalized Error Series

The main disadvantage of defining the steady state error in terms of error constants is that only one of the constant is finite and non-zero for a particular system, where as the other constants are zero or infinity. If any error constant is zero, the steady state error is infinity.

If the inputs are other than step, velocity or acceleration inputs, we can extend the concept of error constants to include the inputs which can be represented by a polynomial. Many functions which are analytic can be denoted by a polynomial in t.

Let the error be given by,

$$E(s)=\frac{R(s)}{1+G(s)}$$

The above equation may be written as,

$$E(s)=Y(s).R(s)$$

Where,

$$Y(s)=\frac{R(s)}{1+G(s)}$$

Using convolution theorem, it can be written as,

$$e(t)=\int_0^t y(\tau)r(t-\tau)d\tau$$

Assuming that r (t) has first n derivatives, r (1- τ) can be expanded into a Taylor series,

$$r(t-\tau)=r(t)-\tau r'(t)+\frac{\tau^2}{2!}r''(t)-\frac{\tau^3}{3!}r'''(t)$$

Where the primes indicates the time derivatives. We have,

$$e(t)=\int_0^t y(\tau)\left[r(t)-\tau r'(t)+\frac{\tau^2}{2!}r''(t)-\frac{\tau^3}{3!}r'''(t)....\right]d\tau$$

$$=r(t)\int_0^t y(\tau)d\tau-r'(t)\int_0^t \tau y(\tau)d\tau-r''(t)\int_0^t \frac{\tau^2}{2!}y(\tau)d\tau+....$$

To obtain the steady state error, we take the limit t →∞on both the sides of equation,

$$e_{ss}=\underset{t\to\infty}{Lt}\, e(t)=\underset{t\to\infty}{Lt}\left[r(t)\int_0^t y(\tau)d\tau-r'(t)\int_0^t \tau y(\tau)d\tau+r''\frac{\tau^2}{2!}y(\tau)d\tau...\right]$$

$$e_{ss} = r_{ss}(t)\int_0^\infty y(\tau)d\tau - r_{ss}'(t)\int_0^\infty \tau y(\tau)d\tau + r_{ss}''(t)\int_0^\infty \frac{\tau^2}{2!}y(\tau)d\tau + \ldots.$$

Where the suffix ss denotes steady state part of the function. It may be further observed that the integrals in equation yield s constant values. Hence it may be written as,

$$e_{ss} = C_0 r_{ss}(t) + C_1 r_{ss}^{*}(t) + \frac{C_2}{2!} r_{ss}^{**}(t) + \ldots.\frac{C_n}{n!} r_{ss}^{(n)}(t)$$

Where,

$$C_o = \int_0^\infty y(\tau)d\tau$$

$$C_1 = -\int_0^\infty \tau y(\tau)d\tau$$

$$C_n = (-1)^n \int_0^\infty \tau^n y(\tau)d\tau$$

The above equation is known as generalized error series.

The coefficients C_0, C_1, C_2,.... C_n are defined as generalized error coefficients. Generalized error series may be observed that the steady state error is obtained as a function of time, in terms of generalized error coefficients, the steady state part of the input and its derivatives. For a given transfer function G(s), the error coefficients can be easily evaluated as shown in the following.

Let,

$$y(t) = \tau^{-1} Y(s)$$

$$Y(s) = \int_0^\infty y(\tau)e^{-st}\, d\tau$$

$$\underset{s\to 0}{\mathrm{Lt}}\, Y(s) = \underset{s\to 0}{\mathrm{Lt}} \int_0^\infty y(\tau)e^{-st}\, d\tau$$

$$= \int_0^\infty y(\tau)\, \underset{s\to 0}{\mathrm{Lt}}\, e^{-st}\, d\tau$$

$$= \int_0^\infty y(\tau)d\tau$$

$$=C_o$$

Taking the derivative of equation with respect to s, we have,

$$\frac{dY(s)}{ds}=\int_{0}^{\infty} y(\tau)(-\tau)e^{-st}d\tau$$

Now taking the limit of equation as $s \to 0$, we have,

$$\underset{s\to 0}{Lt}\frac{dY(s)}{ds}=\int_{0}^{\infty} y(\tau)(-\tau)\ \underset{s\to 0}{Lt}\, e^{-st}d\tau$$

$$=-\int_{0}^{\infty} \tau y\,(\tau)d\tau$$

$$=C_1$$

$$C_2=\underset{s\to 0}{Lt}\frac{d^2\,Y(s)}{ds^2}$$

$$C_3=\underset{s\to 0}{Lt}\frac{d^3\,Y(s)}{ds^3}$$

$$C_a=\underset{s\to 0}{Lt}\frac{d^n\,Y(s)}{ds^n}$$

Thus, the constants can be evaluated using the above equations and the time variation of the steady state error also can be obtained.

Advantages

- It gives the error signal as a function.
- The steady state error can be determined for any type of input by using the generalized error constants.

Problems

Determine the error coefficients for the system having $G(s).H(s)=\dfrac{(s+2)}{s(1+0.5s)(1+0.2s)}$

1.Letusderivetheerrorcoefficientsforthesystemhaving $G(s)H(s)=\dfrac{(s+2)}{s(1+0.5s)(1+0.2s)}$

Solution:

Given:

$$G(s)H(s)=\frac{(s+2)}{s(1+0.5s)(1+0.2s)}$$

Error coefficients are K_p, K_v and K_a.

$$K_p=\lim_{s\to 0} G(s)H(s)=\lim_{s\to 0}\frac{(s+2)}{s(1+0.5s)(1+0.2s)}=\infty$$

$$K_v=\lim_{s\to 0} sG(s)H(s)=\lim_{s\to 0} s\frac{(s+2)}{s(1+0.5s)(1+0.2s)}=\frac{2}{1\times 1}=2$$

$$K_a=\lim_{s\to 0} s^2 G(s)H(s)=\lim_{s\to 0} s^2\frac{(s+2)}{s(1+0.5s)(1+0.2s)}=0$$

2. Let us calculate the minimum value of K_1, if the steady error is to be less than 0.1 and the input r (t) = (1 + 6t) is applied to the unity feedback system that has the forward transfer function $G(s)=\frac{K_1(2s+1)}{s(5s+1)(1+s)^2}$.

Solution:

Given:

Input $r(t)=1+6t$

$$G(s)=\frac{K_1(2s+1)}{s(5s+1)(1+s)^2}$$

On taking Laplace transform of r(t), we get R(s).

$$\therefore\ R(s)=Lr(t)=L1+6t=\frac{1}{s}+\frac{6}{s^2}$$

The error signal in s-domain E(s) is given by,

$$\therefore\ E(s)=\frac{R(s)}{1+G(s)H(s)}=\frac{\frac{1}{s}+\frac{6}{s^2}}{1+\frac{K_1(2s+1)}{s(5s+1)(1+s)^2}}$$

$$\frac{\frac{1}{-} \quad \frac{6}{-}}{\frac{s(5s+1)(1+s) \ +K \ (2s+1)}{s(5s+1)(1+s)}}$$

Here H(s) = 1

$$=\frac{1}{2}\left[\frac{s(5s+1)(1+s)^2}{s(5s+1)(1+s)^2+K_1(2s+1)}\right]+\frac{6}{s^2}\left[\frac{s(5s+1)(1+s)^2}{s(5s+1)(1+s)^2+K_1(2s+1)}\right]$$

The steady state error e_{ss} can be obtained from final value theorem,

$$e_{ss}=\underset{t\to\infty}{Lt}\,e(t)=\underset{s\to 0}{Lt}\,s\,E(s)$$

$$=\underset{s\to 0}{Lt}\,s\left\{\frac{1}{s}\left[\frac{s(5s+1)(1+s)^2}{s(5s+1)(1+s)^{b2}+K_1(2s+1)}\right]+\frac{6}{s^2}\left[\frac{s(5s+1)(1+s)^2}{s(5s+1)(1+s)^2+K_1(2s+1)}\right]\right\}$$

$$=\underset{s\to 0}{Lt}\left\{\frac{s(5s+1)(1+s)^2}{s(5s+1)(1+s)^2+K_1(2s+1)}+\frac{6(5s+1)(1+s)^2}{s(5s+1)(1+s)^2+K_1(2s+1)}\right\}$$

$$=0+\frac{6}{K_1}=\frac{6}{K_1}$$

Given that,

$$e_{ss}<0.1,$$

$$\therefore\ 0.1=\frac{6}{K_1}\ \text{or}\ K_1=\frac{6}{0.1}=60.$$

Another method:

The steady state error for the input, r (t) = 1 + 6t is given by,

$$e_{ss}=\frac{1}{1+K_P}+6\frac{1}{K_V}$$

$$K_p=\frac{\lim}{s\to 0}G(s)H(s)=\frac{\lim}{s\to 0}\frac{K(2s+1)}{s(5s+1)(1+s^2)}\Rightarrow\infty$$

$$K_v = \lim_{s \to 0} sG(s)H(s) = \lim_{s \to 0} s\frac{K}{s(5s+1)(1+s^2)} \Rightarrow K$$

Therefore, steady state error becomes,

$$e_{ss} = \frac{1}{1+K_P} + 6\frac{1}{K_V}$$

$$e_{ss} = \frac{1}{1+\infty} + \frac{6}{K}$$

$$e_{ss} = \frac{6}{K}$$

But, the steady state error limited to 0.8.

$$e_{ss} < 0.1 \Rightarrow \frac{6}{K} < 0.1$$

$$\therefore \; K > \frac{6}{0.1} \Rightarrow K > 60$$

3. Let us determine the steady state temperature for the block diagram that represents a heat treating oven.

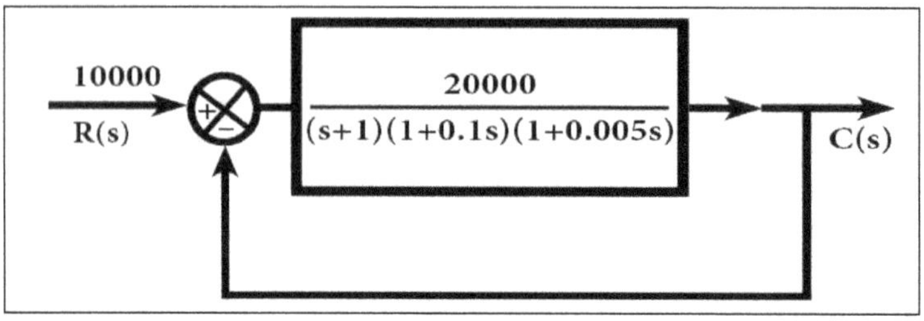

Solution:

The forward path transfer function of the system,

$$G(s) = \frac{20000}{(s+1)(1+0.1s)(1+0.005s)}$$

$H(s) = 1$ and $R(s) = 1000/s$

Since the given system is type '0', therefore the stead y state error,

$$e_{ss} = \frac{A}{1+K_p}, \text{where } K_p = \lim_{s \to 0} G(s)H(s)$$

Position error constant is given by,

$$K_p = \lim_{s \to 0} G(s)H(s) = \lim_{s\to 0} \frac{20000}{(s+1)(1+0.1s)(1+0.005s)} = \frac{20000}{(1+0)(1+0)(1+0)} = 20000$$

Steady state error $e_{ss} = \frac{A}{1+K_p} = \frac{1000}{1+20000} = 0.04999$

Steady state temperature = Desired temperature - Steady state error e_{ss}

$$= 1000 - 0.0499$$

$$= 999.95^{\circ}\ C$$

4. Let us calculate K to limit the error of a system for input $1+8t+\frac{18}{2}t^2$ to 0.8 having $G(s)\,H(s) = \frac{K}{s^2(s+1)(s+4)}$.

Solution:

Given:

Input: $1+8t+\frac{18}{2}t^2$

$$G(s)\,H(s) = \frac{K}{s^2(s+1)(s+4)}$$

The steady state error for the input 1 + 8t + 18 ($t^2/2$) is given by,

$$\text{Steady state error } e_{ss} = \frac{1}{1+K_p} + 8\frac{1}{K_v} + 18\frac{1}{K_a}$$

$$K_p = \lim_{s \to 0} G(s)H(s) = \lim_{s\to 0} \frac{K}{s^2(s+1)(s+4)} \Rightarrow \infty$$

$$K_v = \lim_{s \to 0} sG(s)H(s) = \lim_{s\to 0} s\frac{K}{s^2(s+1)(s+4)} \Rightarrow \infty$$

$$K_a = \lim_{s \to 0} s^2 G(s)H(s) = \lim_{s\to 0} s^2\frac{K}{s^2(s+1)(s+4)} \Rightarrow \frac{K}{4}$$

Therefore, steady state error becomes,

$$e_{ss} = \frac{1}{1+K_p} + 8\frac{1}{K_v} + 18\frac{1}{K_a}$$

$$e_{ss}=\frac{1}{1+\infty}+\frac{8}{\infty}+\frac{18}{K/4}\Rightarrow\frac{72}{K}$$

But the steady state error limited to 0.8.

$$e_{ss}<0.8\Rightarrow\frac{72}{K}<0.8$$

$$K>90$$

5. Let us calculate the open loop transfer function G(s) whose $\frac{C(s)}{R(s)}=\frac{(Ks+b)}{(s^2+as+b)}$.

Solution:

Given:

$$\frac{C(s)}{R(s)}=\frac{(Ks+b)}{(s^2+as+b)}$$

We know that, for a closed loop, the transfer function:

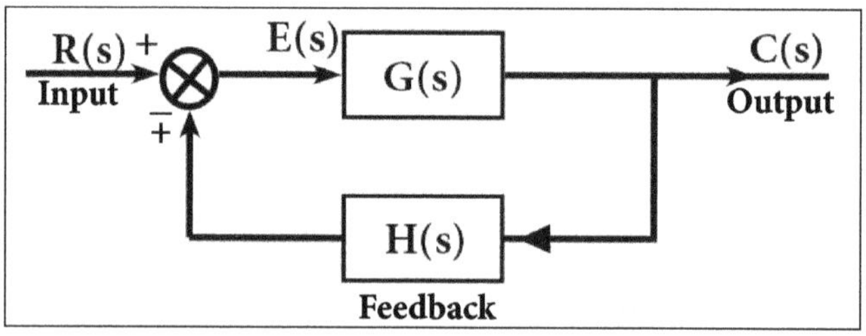

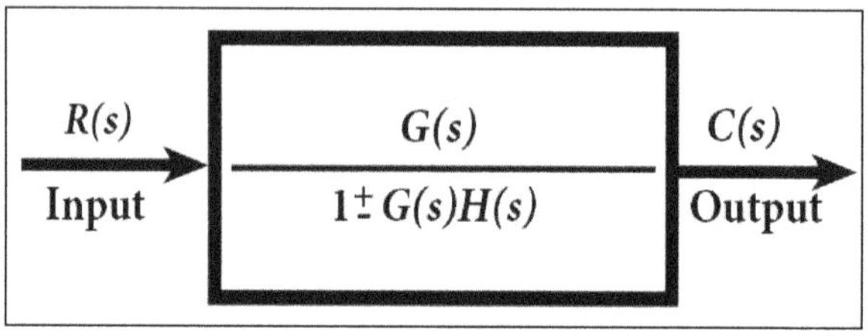

But for a unity feedback system H(s) = 1, then $\frac{C(s)}{R(s)}=\frac{G(s)}{1+G(s)}$

$$\therefore\ \frac{C(s)}{R(s)}=\frac{(Ks+b)}{(s^2+as+b)}=\frac{G(s)}{1+G(s)}\Rightarrow\left[1+G(s)\right]\times(Ks+b)=G(s)\times\left(s^2+as+b\right)$$

$$\Rightarrow Ks+b=G(s)\left[s^2+ab+b-Ks-b\right]$$

$$\therefore\ G(s)=\frac{Ks+b}{s^2+as-Ks}\Rightarrow\frac{Ks+b}{s\left[s+(a-K)\right]}$$

$$\therefore G(s) = \frac{Ks+b}{s\left[s+(a-K)\right]}$$

We know that, the steady state error for a unit ramp input is given by,

$$e_{ss} = \frac{1}{K_v},$$

Where,

$$K_v = \lim_{s \to 0} s.G(s)H(s)$$

$$K_v = \lim_{s \to 0} s \times \frac{Ks+b}{s\left[s+(a-K)\right]} \Rightarrow \frac{b}{(a-K)}$$

Therefore, steady state error $e_{ss} = \frac{(a-K)}{b}$.

6. Let us determine the type of the system, all the error coefficients and error for ramp input with magnitude 4.

Solution:

Given:

The type of the system is 1 (since the number of poles lie at origin = 1).

The error coefficients are K_p, K_v and K_a.

$$K_p = \lim_{s \to 0} G(s)H(s) = \lim_{s \to 0} \frac{40(s+2)}{s(s+1)(s+4)} \Rightarrow \infty$$

$$K_v = \lim_{s \to 0} sG(s)H(s) = \lim_{s \to 0} \frac{40(s+2)}{s(s+1)(s+4)} \Rightarrow \frac{40 \times 2}{1 \times 4} = 20$$

$$K_a = \lim_{s \to 0} s^2 G(s)H(s) = \lim_{s \to 0} s^2 \frac{40(s+2)}{s(s+1)(s+4)} \Rightarrow 0$$

The error for ramp input with magnitude 4 is given by,

$$e_{ss} = \frac{A}{K_v} = \frac{4}{20} = 0.2$$

2.3 Effects of Proportional Derivative and Proportional Integral System

The derivative controller produces a control action based on the rate of change of error signal and it does not produce corrective measures for any constant error. Also derivative controller produces noise. Hence, the derivative controller is not used in control system.

Controllers:

- Proportional controller.
- PD controller.
- PI controller.
- PID controller.

Proportional Controller

The output of the controller is proportional to the value of K, as it varies linearly as the error signal.

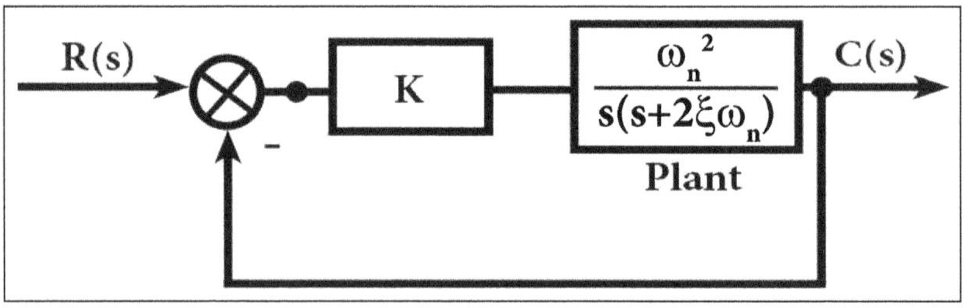

Proportional controller.

PD Controller

The below figure shows a PD controller added to a plant G(s) having unity feedback,

i.e., H(s) = 1

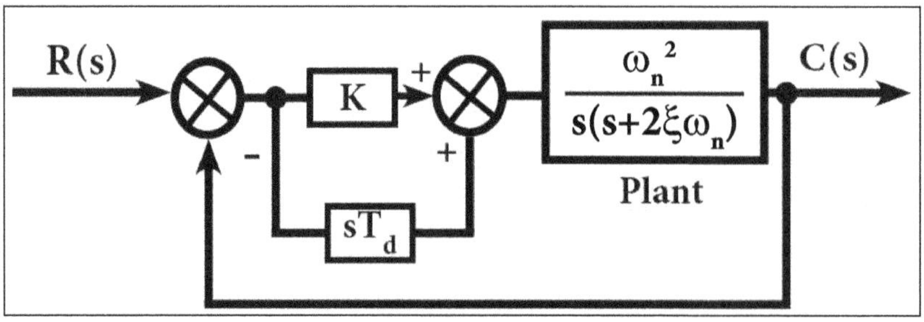

PD controller.

- As there is increase in the damping ratio, peak overshoot and settling time reduces.
- PD controller allows noise to be passed.
- The steady state error for unit step input remains unchanged.

PI Controller

The below figure shows a PI controller added to a plant G(s) having unity feedback. i.e., H(s) = 1.

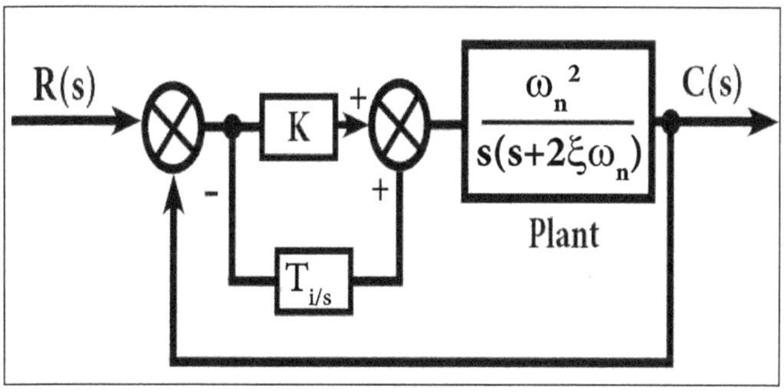

PI controller.

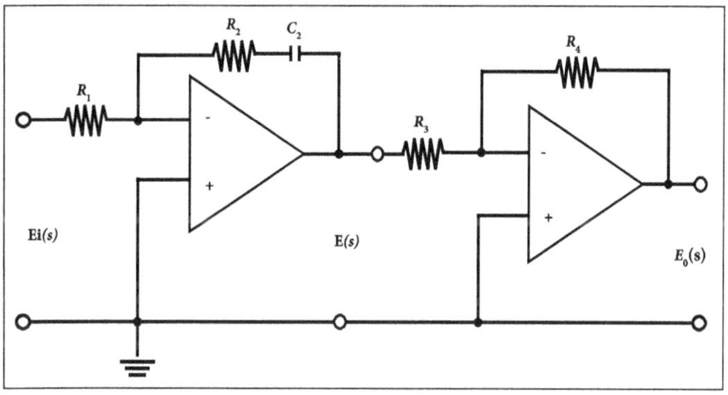

PI controller circuit.

A PI controller affects the system in the following ways:

- Order of the system becomes 3 and type of the system becomes 2, the original is of order 2 and type is 1.
- Since the type of the system increased by 1, error becomes zero for ramp inputs too. i.e., Steady state error of the system improves and the system become more accurate in nature.
- But transient response gets affected if the controller is not designed properly.
- Integral action remains active as long as error is present.

The proportional derivate controller Transfer function is = $K + T_d s$.

The forward path transfer function G(s),

$$G(s)=\frac{(K+T_d s).\omega_n^2}{s^2+2\xi\omega_n s}$$

The closed loop transfer function of the PD controller,

$$\frac{C(s)}{R(s)}=\frac{G(s)}{1+G(s)}=\frac{(K+T_d s).\omega_n^2}{s^2+2\xi\omega_n s+(K+T_d s).\omega_n^2}$$

$$=\frac{(K+T_d s).\omega_n^2}{s^2+(2\xi\omega_n+T_d s\omega_n^2)s+K.\omega_n^2}$$

If ξ is new damping ratio, then,

$$2\xi'\omega_n = 2\xi\omega_n + T_d\omega_n^2$$

$$\xi'=\xi+\frac{\omega_n.T_d}{2}$$

Therefore, damping ratio increases by $\frac{\omega_n T_d}{2}$ and natural frequency remain same.

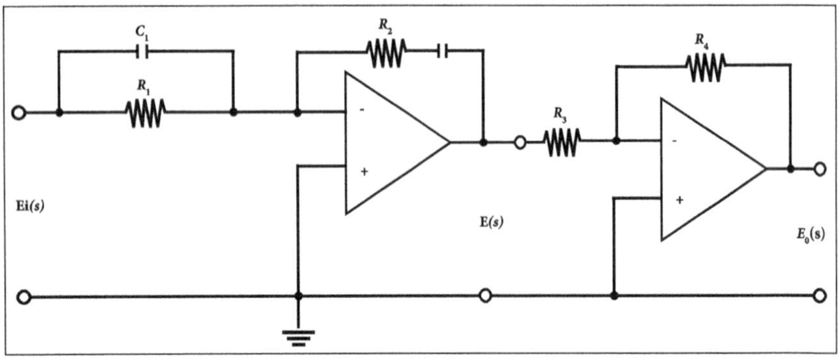

PD controller circuit.

Function of PD Controller

A PD controller affects the system in the following ways:

- Increases the damping ratio.
- Natural frequency of the system remains same.
- Type of the system remains unchanged.

Characteristics of PI controller:

- Infinite gain at zero frequency.
- Increases the type number of the compensated system by 1.

PID Controller

The PID controller used to improve the overall response of the system, where the PD portion of the controller improves the transient response and PI portion of the controller improves the steady state response of the system.

The Transfer Function of PID controller is $\frac{C(s)}{R(s)} = K + sT_d + \frac{T_i}{s}$.

Digital PID Controller

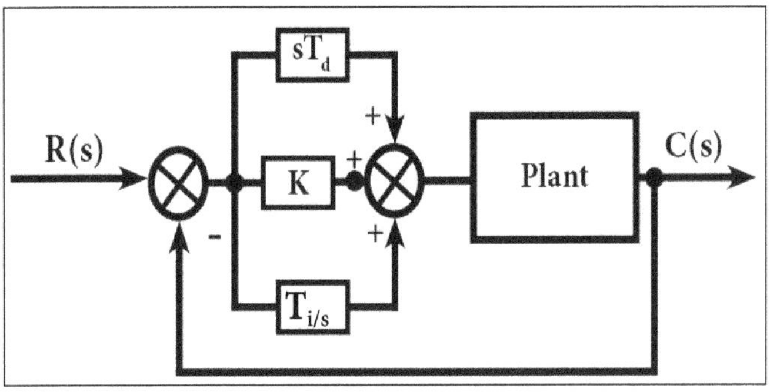

PID controller.

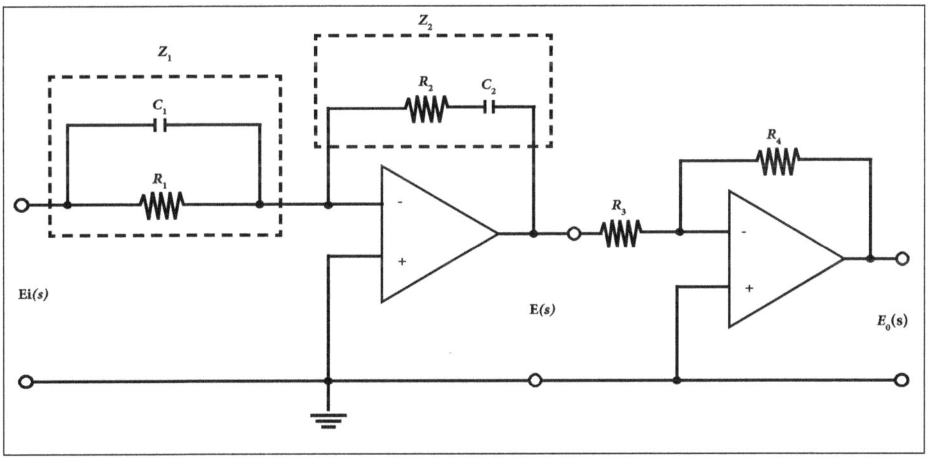

PID controller circuit.

Function of PID Controller

The function of PID controller is to produce an output signal consisting of 3 terms:

- One proportional to the actuating signal.

- Other one proportional to the integral of actuating signal.
- Another one proportional to derivative of actuating signal.

Main job of PID is to improve the transient and steady state performance of the system.

Problems

1. Let us calculate the value of T_d so that the system will be critically d amped and also find its settling time.

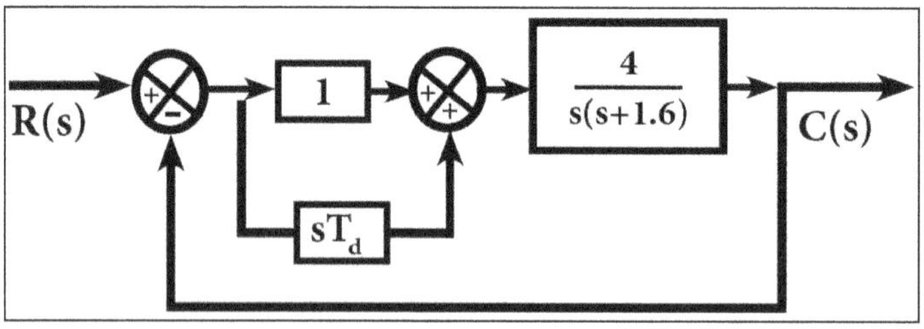

Solution:

Formula to be used:

$$t_s = \frac{4}{\xi'\omega_n}$$

We know that, closed loop transfer function of the system $\frac{C(s)}{R(s)} = \frac{(1+sT_d)4}{s^2+1.6s+4T_ds+4}$

From the above equation, $w_n = 2$ rad / sec.

For critically damped, $\xi' = 1$.

For PD controller, the damping ratio,

$$2\xi' w_n = 1.6+4T_d \Rightarrow \xi' = \frac{1.6+4T_d}{4} \Rightarrow 1 = 0.4+T_d$$

The value of $T_d = 1 - 0.4 = 0.6$

The settling time $t_s = \frac{4}{\xi'\omega_n} = \frac{4}{1X2} = 2$ sec.

3

Stability and Root Locus Technique

3.1 The Concept of Stability

The stability of a system relates to its response to the inputs or disturbances. A system which remains in the constant state unless it is affected by the external action and which returns to a constant state when the external action is removed can be considered as stable. The systems stability can be defined in terms of its response to external impulse inputs.

A system is stable if its impulse response approaches zero as its time approaches to infinity. The system stability can also be defined in terms of the bounded inputs. A system is stable if any bounded input produces the bounded output for all bounded initial conditions.

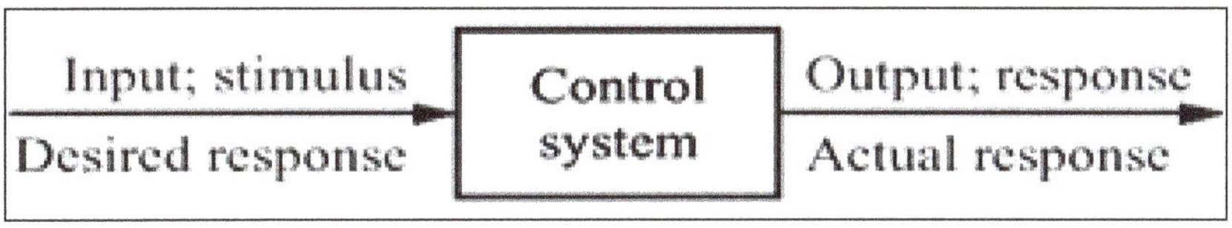

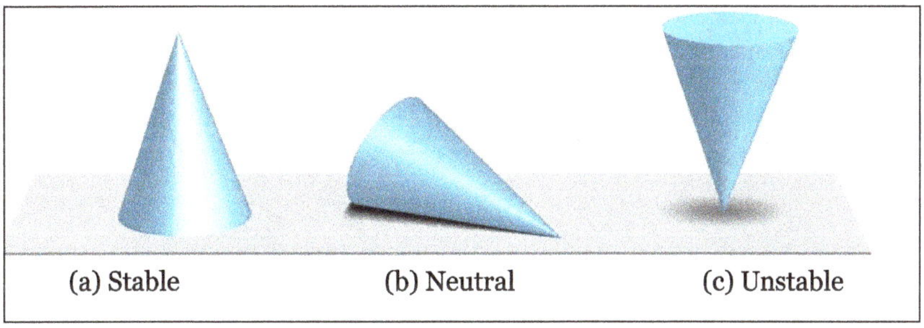

Basic concept of stability.

3.2 Routh's Stability Criterion

Determination of Stability using Routh -Hurwitz Table

Form the Routh Array for the given characteristic equation $1+G(s)\,H(s) = 0$. If all the terms in the first column of the Routh array have positive signs, then the system is said to be stable.

Routh's Criterion

Suppose the characteristic equation of the nth order system is,

$$ao\ S^{n} + a_{1}\ S^{n-1} + a_{2}\ S^{n-2} + \dots + a_{n-1}\ S + ao\ n = 0$$

The Hurwitz determinant is given by,

$$\begin{vmatrix} a_1 & a_0 & 0 & 0 & 0 & 0 & \cdots & 0 & 0 \\ a_3 & a_2 & a_1 & a_0 & 0 & 0 & \cdots & 0 & 0 \\ a_5 & a_4 & a_3 & a_2 & a_1 & a_0 & \cdots & 0 & 0 \\ \vdots & \vdots & \vdots & \vdots & \vdots & \vdots & & \vdots & \vdots \\ a_{2n-1} & a_{2n-2} & a_{2n-3} & \cdots & \cdots & \cdots & \cdots & a_{n+1} & a_n \end{vmatrix}$$

Here the coefficients with Indices larger than n are taken zero. Similarly, the coefficients with negative Indices are replaced with zeros.

The condition of the stability is that n determinants formed from the principal minors of the Hurwitz determinant will be greater than zero. That is,

$$\Delta_1 = a_1 > 0$$

$$\Delta_2 = \begin{vmatrix} a_1 & a_0 \\ a_3 & a_2 \end{vmatrix} > 0$$

$$\Delta_3 = \begin{vmatrix} a_1 & a_0 & 0 \\ a_3 & a_2 & a_1 \\ a_5 & a_4 & a_3 \end{vmatrix} > 0$$

$$\Delta_n = \begin{vmatrix} a_1 & a_0 & 0 & 0 & 0 & 0 & \cdots & 0 & 0 \\ a_3 & a_2 & a_1 & a_0 & 0 & 0 & \cdots & 0 & 0 \\ a_5 & a_4 & a_3 & a_2 & a_1 & a_0 & \cdots & 0 & 0 \\ \vdots & \vdots & \vdots & \vdots & \vdots & \vdots & & \vdots & \vdots \\ a_{2n-1} & a_{2n-2} & a_{2n-3} & \cdots & \cdots & \cdots & \cdots & a_{n+1} & a_n \end{vmatrix} > 0$$

Moreover when $\Delta_{n-1} = 0$, the system is marginally stable.

Problems

1. Let us calculate the stability of the system represented by the characteristic equation $s^4 + 8s^3 + 18s^2 + 16s + 5 = 0$ using Routh criterion.

Solution:

Given:

$$s^4 + 8s^3 + 18s^2 + 16s + 5 = 0$$

s^4	1	18	5
s^3	8	16	0
s^2	$\frac{8*18-1*16}{8}=16$	$\frac{8*5-1*0}{8}$	
s^1	$\frac{16*16-8*5}{16}=13.5$	0	
s^0	5		

From the first column of the two Routh arrays, there is no sign change in the first column. Therefore, the system is stable. All the roots lies L.H.S of the s-plane.

2. Let us find the range of k for the stability of $s^4 + 2s^3 + 2s^2 + (3+K)s + K = 0$, $K > 0$ using Routh criterion.

Solution:

Given:

$$s^4 + 2s^3 + 2s^2 + (3+K)s + K = 0,\ K > 0$$

s^4	1	2	K
s^3	2	$3+K$	0
s^2	$\frac{2*2-1*(3+K)}{2}$	K	
s^1	$\frac{\frac{1-K}{2}*(3+K)-2K}{\frac{1-K}{2}}=\frac{(1-K)(3+K)-4K}{1-K}=\frac{3-6K-K^2}{1-K}$	0	
s^0	K		

For a stab e system, the first column values should be > 0, $\frac{1-K}{2}>0 \Rightarrow K>0.5$

$$\frac{3-6K-K^2}{1-K}>0 \Rightarrow K^2+6K-3=0,\ \text{by solving this,}\ \frac{-3\pm 6.92}{2} \Rightarrow -4.91, 2.41$$

$0.2 < K < 2.41$

3. Let us find the value of K for which the open-loop transfer function of a unity feedback control system is given as $G(s)H(s)=\frac{K}{s(s^2+s+4)}$, which is stable by applying Routh Hurwitz criterion.

Solution:

Given:

$$G(s)H(s)=\frac{K}{s(s^2+s+4)}$$

Formula to be used:

$$1 + G(s)H(s)$$

Routh Hurwitz Criterion

Form the Routh array by taking alternate coefficients of characteristic equation. We know that the characteristic equation 1+G(s) H(s) = 0

$$1+G(s)H(s)=1+\frac{K}{s(s^2+s+4)} \Rightarrow s(s^2+s+4)+K=0$$

$$s^3+s^2+4s+K=0$$

Routh array can be formed as,

$$\begin{array}{lcc} s^3 & 1 & 4 \\ s^2 & 1 & K \\ s^1 & \frac{4-K}{1} & 0 \\ s^0 & K & \end{array}$$

From the Routh array, for stable system, first column must have positive values. Therefore, for a stable system, the range of K value is $0 < K < 4$.

4. Considering the unity feedback system $G(s)=\frac{Ks(s+2)}{(s^2-4s+8)(s+3)}$, let us calculate the range K and also the frequency of oscillation when the system is marginally stable.

Solution:

Given:

$$G(s)=\frac{Ks(s+2)}{(s^2-4s+8)(s+3)}$$

Formula to be used:

$$\frac{C(s)}{R(s)}=\frac{G(s)}{1+G(s)}$$

Characteristic equation 1 + G(s) = 0

Part (i):

First, we find the range of K for stability using the Routh-Hurwitz criterion. The closed-loop transfer function is given by,

$$\frac{C(s)}{R(s)}=\frac{G(s)}{1+G(s)}=\frac{\frac{Ks(s+2)}{(s^2-4s+8)(s+3)}}{1+\frac{Ks(s+2)}{(s^2-4s+8)(s+3)}}$$

$$\frac{C(s)}{R(s)}=\frac{G(s)}{1+G(s)}=\frac{Ks(s+2)}{(s^2-4s+8)(s+3)+Ks(s+2)}$$

Characteristic equation 1 + G(s) = 0

$$(s^2-4s+8)(s+3)+Ks(s+2)=0$$

$$s^3+3s^2-4s^2-12s+8s+24+Ks^2+2Ks=0$$

$$s^3+(K-1)s^2+(2k-4)s+24=0$$

Now we form the Routh table shown in below whose first two rows consist of the coefficients of the denominator:

S^3	1	2K - 4
S^2	K - 1	24
S^1	$\frac{(K-1)(2K-4)-24}{K-1}=\frac{2K^2-6K-20}{K-1}$	0
S^0	24	0

According to the Routh-Hurwitz criterion, the closed-loop system is stable if there are no sign changes in the first column. Hence for stability, first column must be > 0.

$$\frac{2K^2-6K-20}{24}>0 \Rightarrow 2K^2-6K-20>0$$

$$K>5 \text{ or } K<-2$$

Then,

$$K-1>0 \Rightarrow K>1$$

Summarizing, the closed-loop system is stable for K > 5.

Part (ii):

If K = 5, there is an entire row of zeros in the table, which could mean that there are poles on the jω-axis.

Now we continue the analysis of the Routh Table in the special case of a row of zeros. Returning to the s^2 row and replacing K with 5, we form the even polynomial,

$$P(s)=4s^2+24$$

Differentiating with respect to s,

$$\frac{dP}{ds}=8s$$

Replacing the row of zeros with the coefficients of 8, we obtain the Routh table shown as follows:

S^3	1	2*5-4 = 6
S^2	5-1 = 4	24
S^1	8	0
S^0	24	0

Since there are no sign changes from the first column of Routh, the system is marginally stable. Given the frequency of oscillation we solve P(s) = 0, then

$$4s^2+24=0$$

$$s^2=-24/4$$

$$s^2=-6$$

$$s=\pm j\sqrt{6}$$

Hence, the frequency of oscillation is $\omega = \sqrt{6}$ rad/sec.

5. By applying the Routh criterion let us determine the stability of the close d loop system as a function of K for the open loop transfer function of a unity feedback control system $G(s)=\frac{K}{(s+2)(s+4)(s^2+6s+25)}$.

Solution:

Given:

$$G(s)=\frac{K}{(s+2)(s+4)(s^2+6s+25)}$$

Formula to be used:

$$\frac{C(s)}{R(s)}=\frac{G(s)}{1+G(s)}$$

Characteristic equation 1 + G(s) = 0

Step (i):

First, we find the range of K for stability using the Routh-Hurwitz criterion. The closed-loop transfer function is given by,

$$\frac{C(s)}{R(s)}=\frac{G(s)}{1+G(s)}=\frac{\frac{K}{(s^2+6s+25)(s+2)(s+4)}}{1+\frac{K}{(s^2+6s+25)(s+2)(s+4)}}$$

$$\frac{C(s)}{R(s)}=\frac{G(s)}{1+G(s)}=\frac{K}{(s^2+6s+25)(s+2)(s+4)+K}$$

Characteristic equation 1 + G(s) = 0

$$(s^2+6s+25)(s+2)(s+4)+K=0$$

$$(s^3+2s^2+6s^2+12s+25s+50)(s+4)+K=0$$

$$(s^3+8s^2+37s+50)(s+4)+K=0$$

$$s^4 + 4s^3 + 8s^3 + 32s^2 + 37s^2 + 148s + 50s + 200 + K = 0$$

$$s^4 + 12s^3 + 69s^2 + 198s + (200 + K) = 0$$

Step (ii):

Now we form the Routh table shown in below, whose first two rows consist of the coefficients of the above equation:

S^4	1	69	200 + K
S^3	12	198	0
S^2	$\frac{12*69-198}{12}=52.8$	$\frac{12*(200+K)-1*0}{12}=200+K$	0
S^1	$\frac{52.5*198-(200+K)*12}{52.5}=\frac{7995-12\,K}{52.5}$	0	
S^0	200 + K	0	

Step (iii):

According to the Routh-Hurwitz criterion, the closed-loop system is stable if there are no sign changes in the first column. Hence, for stability, first column must be > 0.

$$\frac{7995-12K}{52.5} > 0 \Rightarrow K < \frac{7995}{12} = 666.25$$

and

$$200 + K > 0 \Rightarrow K < -200$$

From the above two conditions, the closed loop system is stable for K < 666.25.

6. Let us find the value of K, which will cause sustained oscillations in the closed loop system and the frequency of sustained oscillation for the system $G(s)=\frac{K}{(s+2)(s^3+10s^2+49s+100)}$. Also using Routh-stability criterion, let us calculate the range of values of K for system to be stable.

Solution:

Given:

$$G(s)=\frac{K}{(s+2)(s^3+10s^2+49s+100)}$$

Formula to be used:

$$\frac{C(s)}{R(s)}=\frac{G(s)}{1+G(s)}$$

Characteristic equation 1 + G(s) = 0

Using Routh-stability criterion,

Step (i):

First, we find the range of K for stability using the Routh-Hurwitz criterion. The closed-loop transfer function is given by,

$$\frac{C(s)}{R(s)}=\frac{G(s)}{1+G(s)}=\frac{\frac{K}{(s^3+10s^2+49s+100)(s+2)}}{1+\frac{K}{(s^3+10s^2+49s+100)(s+2)}}$$

$$\frac{C(s)}{R(s)}=\frac{G(s)}{1+G(s)}=\frac{K}{(s^3+10s^2+49s+100)(s+2)+K}$$

Characteristic equation 1 + G(s) = 0

$$(s^3+10s^2+49s+100)(s+2)+K$$

$$s^4+2s^3+10s^3+20s^2+49s^2+98s+100s+200+K=0$$

$$s^4+12s^3+69s^2+198s+(200+K)=0$$

Step (ii):

Now we form the Routh table shown in table, whose first two rows consist of the coefficients of the above equation:

S^4	1	69	200 + k
S^3	12	198	0
S^2	$\frac{12*69-198}{12}=52.8$	$\frac{12*(200+K)-1*0}{12}=200+K$	0
S^1	$\frac{52.5*198-(200+K)*12}{52.5}=\frac{7995-12K}{52.5}$	0	
S^0	200 + k	0	

Step (iii):

According to the Routh-Hurwitz criterion, the closed-loop system is stable if there are no sign changes in the first column. Hence, for stability, first column must be > 0.

$$\frac{7995-12K}{52.5}>0 \Rightarrow K<\frac{7995}{12}=666.25$$

and

$$200+K>0 \Rightarrow K<-200$$

From the above two conditions, the closed loop system is stable for K<666.25.

Step (iv):

If K = 666.25 will cause sustained oscillation, there is an entire row of zeros in t able, which could mean that there are poles on the jω-axis. Now, we continue the analysis of the Routh Table in the special case of a row of zeros. Returning to the s^2 row and replacing K with 666.25,

$$P(s)= 52.5s^2+200+K$$

Step (v):

Given the frequency of oscillation we solve P(s) = 0,

Then,

$$52.5s^2+200+666.25=0$$

$$52.5s^2+866.25=0$$

$$s^2=-866.25/52.5$$

$$s^2=-16.5$$

$$s=\pm j\sqrt{16.5}$$

Hence, the frequency of oscillation is $\omega = \sqrt{16.5}$ rad/sec.

7. Let us investigate the stability of a unity feedback control system whose open-loop transfer function is given by $G(s)=\frac{e^{-st}}{s(s+2)}$.

Solution:

Given:

$$G(s)=\frac{e^{-st}}{s(s+2)}$$

Formula to be used:

$$\frac{C(s)}{R(s)}=\frac{G(s)}{1+G(s)}$$

Characteristic equation 1 + G(s) = 0

Step (i):

First, we find the range of K for stability using the Routh-Hurwitz criterion. The closed-loop transfer function is given by,

$$\frac{C(s)}{R(s)}=\frac{G(s)}{1+G(s)}=\frac{\frac{e^{-st}}{s(s+2)}}{1+\frac{e^{-st}}{s(s+2)}}$$

$$\frac{C(s)}{R(s)}=\frac{G(s)}{1+G(s)}=\frac{e^{-st}}{s(s+2)+e^{-st}}$$

Characteristic equation 1 + G(s) = 0

$$s(s+2)+e^{-st}=0$$

$$s^2+2s+(1-st)=0$$

$$s^2+(2-t)s+1=0$$

Step (ii):

Now we form the Routh table shown in table, whose first two rows consist of the coefficients of the above equation:

S^2	1	1
S^1	2 - t	0
S^0	1	0

Step (iii):

According to the Routh-Hurwitz criterion, the closed-loop system is stable if there are no sign changes in the first column. Hence for stability, first column must be > 0.

$$2-t>0 \Rightarrow t<2$$

Therefore, the system is stable for t < 2.

8. Let us form the Routh array of the system represented by the characteristic polynomial equation $s^6 + 2s^5 + 8s^4 + 12s^3 + 20s^2 + 16s + 16 = 0$.

Solution:

Given:

$$s^6 + 2s^5 + 8s^4 + 12s^3 + 20s^2 + 16s + 16 = 0$$

Step (i):

Now we form the Routh table shown in Table, whose first two rows consist of the coefficients of the above equation:

s^6	1	8	20	16
s^5	2	12	16	0
s^4	$\frac{2*8-1*12}{2}=2$	$\frac{2*20-1*16}{2}=12$	$\frac{2*16-1*0}{2}=16$	
s^3	$\frac{2*12-2*12}{2}=0$	$\frac{2*16-2*16}{2}=0$	0	
s^2				
s^1				
s^0				

Step (ii):

Since the entire row of s^3 is zero, then by taking (i.e, s^4 row) Aux Polynomial equation and differentiating with respect to s.

$$P(S) = 2s^4 + 12s^2 + 16$$

$$P(S) = s^4 + 6s^2 + 8$$

Differentiate with respect to 's'

$$\frac{dP}{ds} = 4s^3 + 12s$$

Step (iii):

Replace the above row of s^3 in Routh Array by the coefficients of dP/ds, then:

s^6	1	8	20	16
s^5	2 or 1	12 or 6	16 or 8	0
s^4	2 or 1	12 or 6	16 or 8	
s^3	4	12	0	
s^2	$\frac{4*6-1*12}{4}=\frac{24-12}{4}=3$	$\frac{4*8-1*0}{4}=8$		
s^1	$\frac{3*12-4*8}{3}=\frac{36-32}{3}=\frac{4}{3}$	0		
s^0	8			

Step (iv):

According to the Routh-Hurwitz criterion, the closed-loop system is stable if there are no sign changes in the first column. From the 1st column of the Routh Array, all the coefficients are positive. So, all the roots of the equation lies in left side of the s-plane. Therefore, the system is stable.

9. Consider that a unity feedback system is characterized by the open-loop transfer function $G(s)=\frac{k}{(s+2)(s^3+10s^2+49s+100)}$. Using Routh-Stability criterion, let us calculate the range of values of K for system to be stable. Also determine the value of K, which will cause sustained oscillations in the closed loop system and also the frequency of sustained oscillations.

Solution:

Given:

$$G(s)=\frac{k}{(s+2)(s^3+10s^2+49s+100)}$$

Formula to be used:

$$\frac{C(s)}{R(s)} = \frac{G(s)}{1+G(s)H(s)}$$

The closed loop transfer function is given by,

$$\frac{C(s)}{R(s)} = \frac{G(s)}{1+G(s)H(s)}$$

$$= \frac{\dfrac{k}{(s+2)(s^3+10s^2+49s+100)}}{1+\dfrac{k}{(s+2)(s^3+10s^2+49s+100)}}$$

$$= \frac{k}{(s+2)(s^3+10s^2+49s+100)+k}$$

$$\frac{C(s)}{R(s)} = \frac{k}{s^4+12s^3+69s^2+198s+200+k}$$

The characteristics equation is given by,

$$s^4+123+69s^2+19.8s+200+k=0$$

The Routh array is constructed as shown below:

s^4	1	69	200 + k
s^3		12	198

Divide s^3 row by 12 to simplify the calculations:

s^4	1	69	200 + k
s^3	1	16.5	
s^2	52.5	200 + k	
s^1	$\frac{666.25-k}{52.5}$		
s^0	200 + k		

For the system to be stable there should not be any sign change in first column. Hence choose k so that first column is positive.

From s^1 row, f or stable system 666.25 - k = 0.

The k should be less than 666.25.

From s^0 row, for stable system (200 + k) = 0.

Since k must be greater than -200, but practical value of k starts from 0.

Therefore range of k is $0 < k < 666.25$.

When k = 666.25, s row becomes zero. Thus, the co-efficient of auxiliary equation are given by,

$$52.53s^2 + 200 + k = 0$$

$$52.53s^2 + 200 + 666.25 = 0$$

$$S^2 = -\frac{200 - 666.25}{52.5} = -16.5$$

$$S = \pm j\sqrt{16.5} = \pm j4.06.$$

When K = 666.25, the system has roots on imaginary axis and so it oscillates. The frequency of oscillation is given by the value of root on imaginary axis.

$$\omega = 4.06 \text{ rad/sec.}$$

3.2.1 Relative Stability Analysis

- In practice, it is desired to determine the relative stability.
- The relative stability of a system can be defined as the property which can be measured by the relative real part of each root or pair of roots.
- Because the relative stability of a system is dictated by location of the roots of the characteristic equation, we can extend the Routh-Hurwitz criterion to ascertain the relative stability.
- This can be accomplished by utilizing the change of variable s, which shifts the s-plane vertical axis to utilize the Routh-Hurwitz criterion.
- For the correct magnitude of shift the vertical axis must be obtained on a trial-and-error basis.
- One may determine the real part of the dominant roots without solving the higher order polynomial q(s).

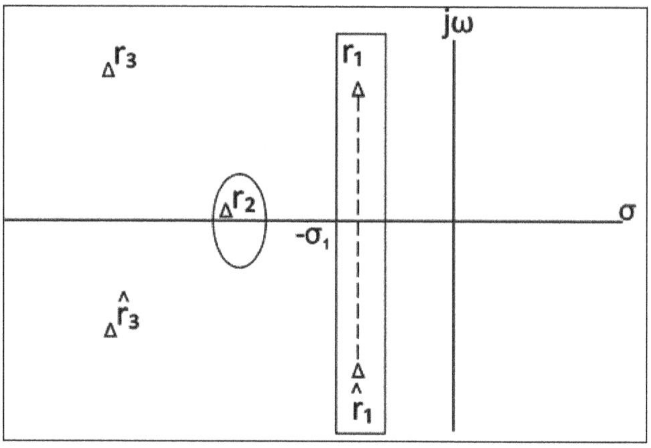

Example: Axis Shift.

Consider the simple third-order characteristic equation,

$$q(s) = s^3 + 4s^2 + 6s + 4$$

First try: Let $s_n = s + 2$. There is no zero in the first column of Routh array, but with two unstable roots.

Second try: $s_n = s + 1$. We have,

$$(s_n - 1)^3 + 4(s_n - 1)^2 + 6(s_n - 1) + 4 = s_n^3 + s_n^2 + s_n + 1.$$

s_n^3	1	1
s_n^2	1	1
s_n^1	0	0
s_n^0	1	0

$$(s_n) = s_n^2 + 1 = (s_n + j)(s_n - j) = (s + 1 + j)(s + 1 - j).$$

The shifting of the s-plane axis to ascertain the relative stability of the system is a very useful approach, particularly for the higher-order system with several pairs of closed-loop complex conjugate roots.

3.2.2 More on the Routh Stability Criterion

Routh stability criterion is also known as modified Hurwitz criterion of stability of the system.

Necessary and sufficient condition for stability in Routh stability criterion:

- The necessary and sufficient condition for stability is that all of the elements in the first column of the Routh array be positive. If this condition is not met, the system is unstable.

- The number of sign changes in the elements of first column of the Routh array corresponds to the number of roots of the characteristic equation in the right half of s-plane.

Difficulties in applying Routh-Hurwitz criterion:

- It is valid only for real coefficients of the characteristic equation.
- It does not provide the exact locations of the closed loop poles in left or right half of s-plane.
- It does not suggest methods of stabilizing an unstable system.
- Applicable only to linear system.

Advantages of Routh's array method:

- Stability of the system can be judged without actually solving the characteristic equation.
- No evaluation of determinants, which saves calculation time.
- For unstable system, it gives number of roots of characteristic equation having positive real root.
- Relative stability of the system can be easily judged.

3.3 Root Locus Concept

The root locus is a way of presenting the graphical information about a system's behavior when the controller is working. Root locus is the widely used tool for the design of closed loop systems and it has the virtue of being a good design tool for continuous time systems and for sampled systems.

The relative stability and the transient performance of the closed loop system are directly related to the location of closed-loop roots of the characteristic equation in the s - plane It is necessary to adjust one or more system parameters to obtain suitable root location. It is useful to determine the locus of roots in s-plane as the parameter varied since the roots are the function of system's parameter.

3.3.1 The Root Locus Concepts

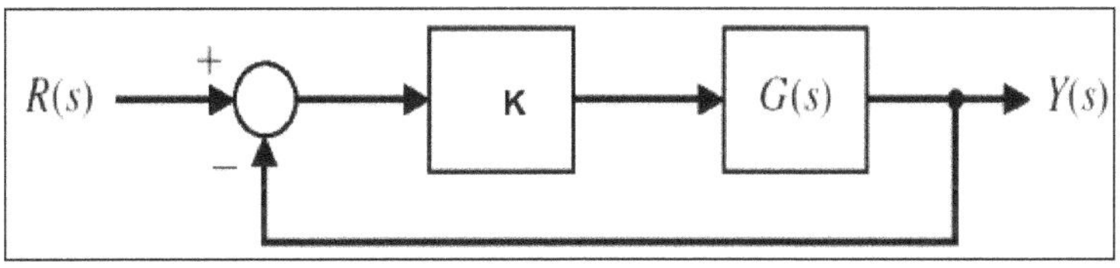

Characteristic equation:

$$T=\frac{KG(s)}{1+KG(s)}$$

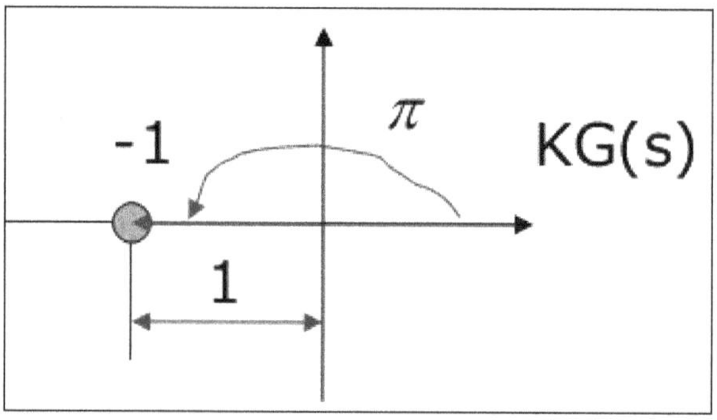

$1 + KG(s) = 0$

$KG(s) = -1$

$KG(s) = -1 + j0$ (Cartesian form)

$|KG(s)| \angle KG(s) = e^{-j\pi}$ (polar Form)

The values of s that fulfill the angle and magnitude conditions are given by,

$$|KG(s)| = 1 \qquad \angle KG(s) = \pm k360° + \pi$$

The roots of the characteristic equation or the closed- loop poles.

$$1 + KG(S) = 0$$

The root locus is the path of the roots of the characteristic equation traced out in the s-plane as a system parameter (K) is changed. ($0 < K < \infty$).

$$1+KG(s) = 0$$

$$1+K\frac{N(s)}{D(s)}=0$$

Where $G(s) = N(s)/D(s)$

$$D(s) + KN(s) = 0$$

When K=0, this collapses to D(s) = 0. Since the roots of D(s) = 0 are the poles of G(s), they are the closed-loop poles for K=0.

When K is large,

$$D(s)+KN(s)=\frac{1}{K}+\frac{N(s)}{D(s)}=0 \Rightarrow \frac{N(s)}{D(s)}=0$$

Thus the closed-loop poles tend to the roots of N(s) = 0, i.e. , the open-loop zeros.

If N(s)/D(s) is strictly proper, the closed-loop poles tend to infinity.

$$\frac{N(s)}{D(s)}=0 \quad \text{if } D(s)\rightarrow\infty$$

Procedure to Plot Root Locus

- Determine all the roots and poles from the open loop transfer function and then plot them on the complex plane.
- All the root loci starts from the poles where k = 0 and terminates at the zeros where K tends to infinity. The number of branches terminating at infinity equals to the difference between the number of poles & number of zeros of G(s) H(s).
- Find the region of existence of root loci.
- Calculate the breakaway points and break in points if any.
- Plot the asymptotes and centroid point s on the complex plane for the root loci by calculating the slope of the asymptotes.
- Now let us calculate the angle of departure and the intersection of root loci with imaginary axis.
- Now let us determine the value of K.
- By following the above procedure, we can easily draw the root locus plot for any open loop transfer function.
- Calculate the gain margin.
- Calculate the phase margin.

3.3.2 Construction of Root Locus

Rules in designing a root locus,

Rule 1: Symmetry:

As all roots are either real or complex conjugate pairs so that the root locus is symmetrical to the real axis.

Rule 2: Number of Branches:

The number of branches of the root locus is equal to the number of poles of the op en-loop transfer function.

Rule 3: Locus Start and End Points:

- The locus starting points ( ) are at the open-loop poles.
- The locus ending points ( ) are at the open-loop zeros.
- Branches end at infinity.
- Number of starting branches from a pol e and ending branches at a zero is equal to the multiplicity of the poles and zeros respectively.

Rule 4: Real Axis Locus:

If the total number of poles and zeros to the right of the point on the real axis is odd, this point lies on the locus.

Rule 5:

The n – m root locus branches that tends to infinity. So along straight line, asymptotes makes an angle given by,

$$\varphi_A = \frac{180^\circ(2g+1)}{n-m};\ q=0,1,0,\ldots,n-m.$$

Rule 6:

The point of intersection of asymptotes with real axis is given by,

$$\sigma_A = \frac{(\text{Sum of poles} - \text{Sum of Zero})}{n-m}$$

Rule 7:

The breakaway and break-in points of root locus are determined from the roots of the equation $dk/ds = 0$.

Rule 8:

The angle of departure from complex open loop pole is given by,

$$P = \pm 180^\circ(2g + 1);\ g = 0,1,2,\ldots$$

The angle of arrival from complex open loop zero is given by,

$$Z = \pm 180^\circ(2g + 1);\ g = 0,1,2,\ldots$$

Rule 9:

The point of interaction of root locus branches with imaginary axis can be determined by using Routh criterion.

Rule 10:

The open loop gain k a t any point on root locus is given by,

$$k = \frac{\text{Product of vector length from open loop poles}}{\text{Product of vector length from open loop zeros}}$$

Advantages of Root Locus Technique

- Root locus technique in control system is easy to implement as compared to other methods.
- With the help of root locus, we can easily predict the performance of the whole system.
- Root locus provides the better way to indicate the parameters.

Root Locus Procedure

Step 1: Write the characteristic equation as 1 + F(s) = 0

Step 2: Rewrite preceding equation into the form of poles and zeros as follows,

$$1 + K\frac{\prod_{j=1}^{m}(s - z_j)}{\prod_{i=1}^{n}(s - p_i)} = 0$$

Step 3: Locate the poles and zeros with specific symbols, the root locus begins at the open-loop poles and ends at the open loop zeros as K increases from 0 to infinity.

If open-loop system has n-m zeros at infinity, there will be n-m branches of the root locus approaching the n-m zeros at infinity.

Step 4: The root locus on the real axis lies in a section of the real axis to the left of an odd number of real poles and zeros.

Step 5: The number of separate loci is equal to the number of open-loop poles.

Step 6: The root loci must be continuous and symmetrical with respect to the horizontal real axis.

Step 7: The loci proceed to zeros at infinity along asymptotes centered at centroid and with Angles.

$$\sigma_a = \frac{\sum_{i=1}^{n} p_i - \sum_{j=1}^{m} z_j}{}$$

$$\phi_a = \frac{(2k+1)\pi}{n-m} \quad (k=0,1,2,\cdots n-m-1)$$

Step 8: The actual point at which the root locus crosses the imaginary axis is readily evaluated by using Routh's criterion.

Step 9: Determine the break-away point d (usually on the real axis).

Step 10: Plot the root locus that satisfy the phase criterion.

$$\angle P(s)=(2k+1)\pi \qquad k=1,2,\cdots$$

Step 11: Determine the parameter value K_1 at a specific root using the magnitude criterion.

$$K_1 = \left. \frac{\prod_{i=1}^{n}(s-p_i)}{\prod_{j=1}^{m}(s-z_i)} \right|_{s=s_1}$$

Problems

1. Let us sketch the root locus for the system: $G(s)=\frac{k(s+3)}{s(s+1)(s+2)(s+4)}$

Solution:

Given:

$$G(s)=\frac{k(s+3)}{s(s+1)(s+2)(s+4)}$$

Formula to be used:

$$\text{Angle of asymptotes} = \pm\frac{180°(2q+1)}{n-m}$$

Step 1: Poles are at s = 0, −1, −2, − 4 (∴ n = 4).

Zeros are at s = − 3. (∴ m = 1).

Step 2: The segment of real axis between s = 0 and s = −1 and from s = −2 to s =−3 and from s = −4 to s = - ∞ will be part of root locus.

Step 3: Angle of asymptotes $=\pm\dfrac{180^\circ(2q+1)}{n-m}$

Here,

$$n-m = 4 - 1 = 3$$

$$q=0, \text{Angles}=\frac{\pm 180^\circ}{3}=\pm 60^\circ$$

$$q=1, \text{Angles}=\frac{\pm 180^\circ(2+1)}{3}=\pm 180^\circ$$

$$q=2, \text{Angles}=\frac{\pm 180^\circ\times 5}{3}=\pm 300^\circ=-+60^\circ$$

$$\text{Centroid}=\frac{\sum P-\sum Z}{A-m}=\frac{(0-1-2-4)-(-3)}{3}$$

$$\sigma=-\ 1.33$$

Step 4: The characteristic equation is given by,

$$1+G(s)H(s)=0$$

$$1+K(s+3)=0$$

$$s(s+1)\ (s+2)\ (s+4)$$

$$\Rightarrow s(s+1)\ (s+2)\ (s+4)+k(s+3)=0.$$

$$K=-\frac{s(s+1)\ (s+2)\ (s+4)}{(s+3)}$$

$$=\frac{-s(s+1)(s^2+6s+8)}{(s+3)}$$

$$=\frac{-(s^2+s)(s^2+6s+8)}{(s+3)}\Rightarrow\frac{\left[s^4+7s^3+14s^2+8s\right]}{(s+3)}$$

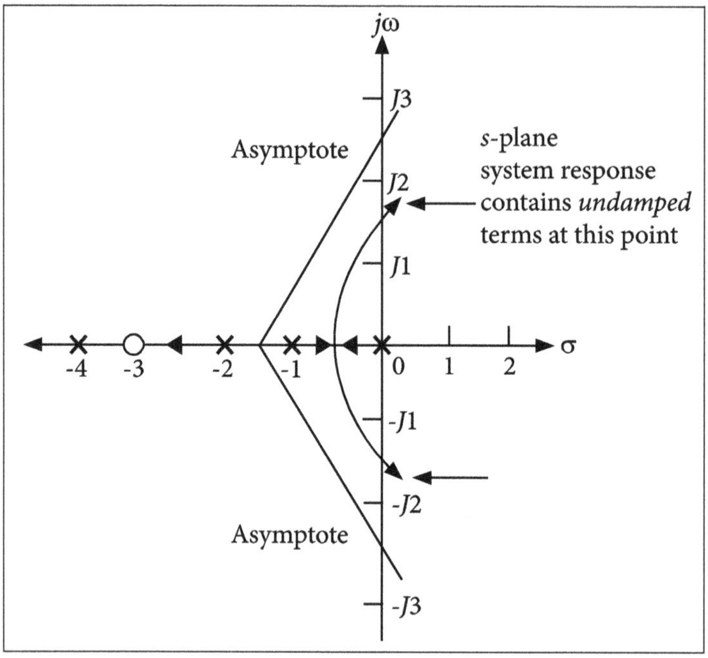

Graph

$$\Rightarrow \frac{-\left[4s^3+21s^2+28s+8\right](s+3)-\left[-(s^4+7s^3+14s^2+8s)(1)\right]}{(s+3)^2}$$

$$\Rightarrow \frac{-(3s^4+26s^3+77s^2+84s+24)}{(s+3)^2}$$

$$\frac{dk}{ds}=0 \Rightarrow 3s^4+26s^3+77s^2+84s+24=0$$

$$\Rightarrow s^4+8.67s^3+25.6s^2+28s+8=0$$

$$\Rightarrow s^4+8.67s^3+25.6s^2+28s+8=0.$$

At s = 0.43, the break-away point exists.

Step 5: Since there are no complex pole or zero, there is no angle of departure or arrival.

Step 6: Crossing point of imaginary axes.

Characteristic equation is given by,

$$s(s+1)(s+2)(s+4)+k(s+3)=0.$$

$$\Rightarrow s^4+7s^3+14s^2+8s+ks+3k=0.$$

Put,

$$s = j\omega,$$

$$(j\omega)^4 + 7(j\omega)^3 + 14(j\omega)^2 + 8(j\omega) + k(j\omega) + 3k = 0.$$

$$\omega^4 - 7j\omega^3 - 14\omega^2 + 8j\omega + kj\omega + 3k = 0.$$

Imaginary part	Real part
$-7j\omega^3 + 8j\omega + kj\omega = 0$ $7\omega^3 = -8\omega - k\omega$ $\omega^2 = \frac{8+k}{7}$	$\omega^4 - 14\omega^2 + 3k = 0.$ $\left(\frac{8+k}{7}\right)^2 - 14\left(\frac{8+k}{7}\right) + 3k = 0$ $\frac{(8+k)^2 - 14\times7(8+k) + 3\times7^2 k}{7^2} = 0$ $(8+k)^2 - 98(8+k) + 147k = 0.$
	$64 + k^2 + 16k - 784 - 98k + 147k = 0.$ $k^2 + 65k - 720 = 0.$ $k = \frac{-65 \pm \sqrt{65^2 + 4\times1\times720}}{2}$ $= 9.6 - 74.6$

The value of k = – 74.6 is neglected as it is a negative value.

$$\therefore\ \omega^2 = \frac{8+k}{7} = \frac{8+9.6}{7} = 2.5$$

$$\omega = \pm\sqrt{2.5} = \pm 1.58$$

$\omega = \pm j1.58$ is the crossing point of root locus.

The value of k at this crossing point is k = 9.6.

2. F or the system shown below let us determine the frequency & gain, K, for which the root locus crosses the imaginary axis. Also determine the range of K.

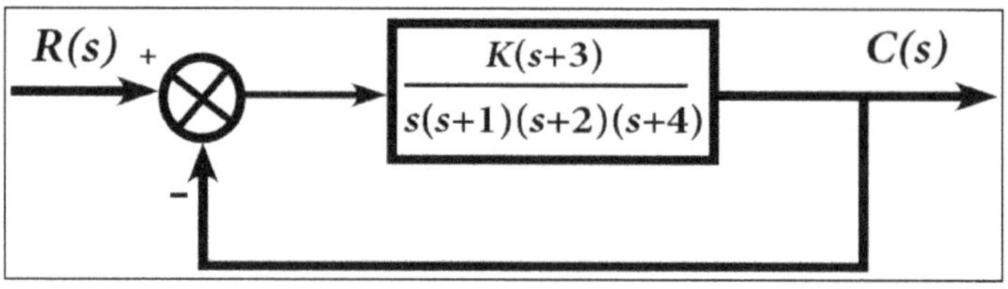

Solution:

Formula to be used:

$$\text{C.L.T.F of } T(s) = \frac{G(s)}{1+G(s)H(s)}$$

$$\text{C.L.T.F of } T(s) = \frac{G(s)}{1+G(s)H(s)},\ H(s)=1$$

Where,

$$T(s) = \frac{K(s+3)}{s^4+7s^3+14s^2+(8+K)s+3K}$$

Form Routh table,

S^4	1	14	3K
S^3	7	8 + K	
S^2	90 - K	21K	
S^1	$\frac{-K^2-65K+720}{90-K}$		
S^0	21K		

For +ve K, only s1 row can be all zeros.

Let,

$-K^2 - 65K + 720/90 - K = 0$ to find value of K on jω -axis.

$-K^2 - 65K + 720 = 0$ $K = 9.65$

To determine the frequency on the jω axis crossing, form the even polynomial by using the s^2 row and with $K= 9.65$,

$$(90-K)s^2+21K=0$$

$$80.35s^2+202.7=0$$

$$s^2 = -202.7/80.35$$

$$s = +\,j1.59$$

The root-locus crosses the jω axis at + j1.59 at a gain of 9.65

Thus, the system is stable for 0 < K < 9.65

3. Given the unity feedback system, let us determine the angle of departure from the complex poles & sketch the root locus.

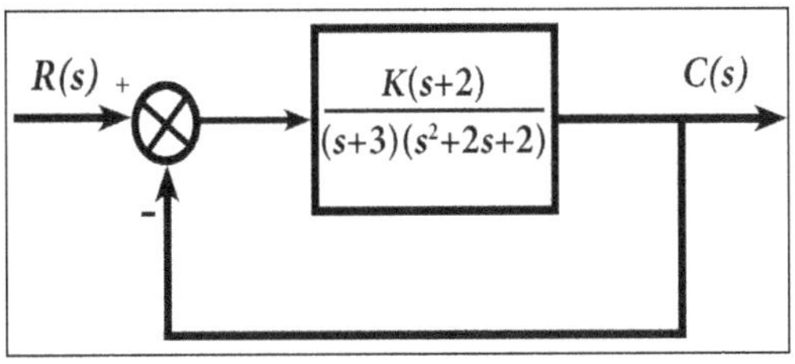

Solution:

$$KG(s)H(s) = \frac{K(s+2)}{(s+3)(S^2+2s+2)}, H(s)=1$$

Where,

$$KG(s)H(s) = \frac{K(s+2)}{(s+3)(s+1-j1)(s+1+j1)}$$

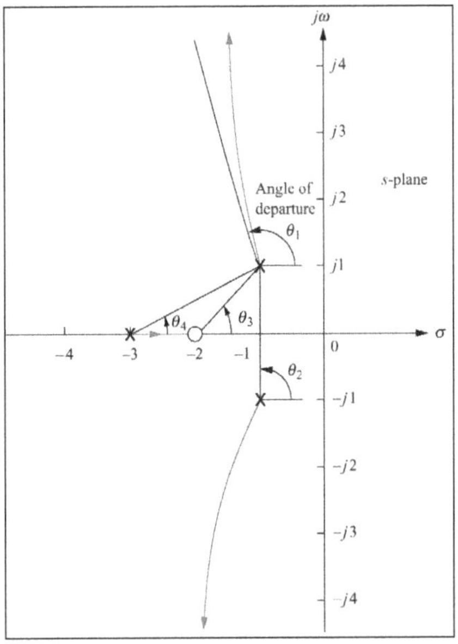

Root locus for the system showing angle of departure.

$$-\theta_1 - \theta_2 + \theta_3 - \theta_4 = (2k+1)180° = 180° (k=0)$$

$$-\theta_1 - 90° + \tan^{-1}\left(\frac{1}{1}\right) - \tan^{-1}\left(\frac{1}{2}\right) = 180°$$

$$\theta_1 = -90° + 45° - 26.5° - 180°$$

$$= -251.6° = 108.4°$$

The angle of departure of the complex pole is -108.4° (symmetry about the real axis).

4

Frequency Response Analysis

4.1 Introduction to Frequency Domain Specifications

When a linear system is subjected to sinusoidal input perturbation, its ultimate response after a long time also becomes a sinusoidal wave, however with different amplitude and phase shift. This is the basis of frequency response analysis.

Frequency Domain Specifications

Resonant Peak (μ_r): The maximum value of the magnitude of closed loop transfer function is called resonant peak.

Resonant Frequency (μ_f): The frequency at which the resonant peak occurs is called resonant frequency.

Bandwidth: The bandwidth is the range of frequencies for which the system gain is more than 3 d B. The bandwidth is a measure of the ability of a feedback system to reproduce the input signal, noise rejection characteristics and rise time.

Cut-off rate: The slope of the log-magnitude curve near the cut-off is called cut-off rate. The cut-off rate indicates the ability to distinguish the signal from noise.

Gain Margin: The gain margin, kg is defined as the reciprocal of the magnitude of the open loop transfer function at phase cross over frequency.

$$\text{Gain margin kg} = 1/\left|G\left(j\omega_{pc}\right)\right|.$$

Phase Cross Over: The frequency at which the phase of the open loop transfer functions is termed as phase cross over frequency ω_{pc}.

Phase Margin: The phase margin γ is the amount of phase lag at the gain cross over frequency (ω_{gc}) which is required to bring the system to the verge of instability.

Gain Cross Over: The gain cross over frequency, ω_{gc} is the frequency at which the magnitude of the open loop transfer f unction is unity.

Advantages of Frequency Response Analysis:

- Used to find the absolute and relative stability of the closed loop system.

- The practical testing of systems can be easily carried out with available sinusoidal signal generators and precis e measurement equipment's.
- Used to find transfer function of complicated systems.
- Easy to carry out the design and parameter adjustment of the open loop transfer function of a system for specified closed loop performance.
- The effects of noise disturbance and parameter variations are easy to visualize and incorporate corrective measures.
- It can be extended to certain non-linear control systems.

4.2 Bode Diagrams

Bode plots are the most widely used means of displaying and communicating frequency response information. It is the graphical representation of a linear, time-invariant system transfer function. In the linear system, any sinusoidal signal that inputs the system only changes in magnitude, when it is amplified and phase, when delayed. Therefore, the system can be described for every frequency, just by its gain and phase shift.

The bode plots simply traces the gain and phase shift of the system to a range of frequencies. There are two bode plots, one plotting the magnitude (or gain) versus frequency and another plotting the phase versus frequency. They are represented with frequency in the logarithmic scale, magnitude in decibels and phase in a linear scale.

There are two ways of drawing a bode plot. One is taking the magnitude and phase of the system transfer function at each frequency and drawing the plot with those points. The other, called asymptotic bode plot, considers straight lines between poles or zeros and has some simple rules for the slopes of those lines.

In the bode plots, commonly encountered frequency responses have a simple shape which means that the laboratory measurements can easily be discerned to have the common factors that lead to those shapes.

Asymptotic Drawing

Bode plot is a standard format for plotting the frequency response of LTI systems. This format is useful because:

- Many common system behaviors produce simple shapes on the Bode plot, so it is easy to either look at a plot and recognize the system behavior or to sketch the plot from the system behavior.

- It is a standard format, so using that format facilitates the communication between engineers.

For example, first order systems have two straight line asymptotes and if we take the data and plot a Bode plot from those data, we can pick out first order factors in a transfer function from the straight line asymptotes.

The format is a log frequency scale on the horizontal axis and on the vertical axis, phase in degrees and magnitude in decibels.

Decibels

For voltages or other physical variables such as current, velocity, pressure, etc.,

$$\text{Decibels }(\text{dB}) = 20\ \log_{10} V_{out} / V_{in}$$

It can be rewritten in terms of power as,

$$\text{Decibels }(\text{dB}) = 20\ \log_{10} V_{out} / V_{in}$$

$$= 10\ \log_{10}(V_{out} / V_{in})^2$$

$$= 10\ \log_{10}(P_{out} / P_{in})$$

Common Values

Common values used are as follows:

$$10\ \log_{10} 2 = 20\ \log_{10} 2 = 3\ \text{dB}$$

$$10\ \log_{10} 1/2 = 20\ \log_{10} 1/\sqrt{2} = -3\ \text{dB “half power”}$$

$$10\ \log_{10} 10 = 20\ \log_{10} 10 = 10\ \text{dB}$$

$10\ \log_{10}\ 100 = 20$ log10 10 = 20 dB, etc 10 dB for every factor of 10 in power.

Procedure for bode plot:

- Convert the given transfer function G(s) into Bo de form by substituting s = jw.
- Prepare a table like below:

Term	Corner Frequency	Slope	Change in slope

- Choose lowest and highest corner frequencies w_L, w_H.
- Calculate magnitude at every corner frequency by using the formula,

$$|M|_{wcl} = \text{Change in slope from } W_{cl} \text{ to } W_L * \log\left(\frac{W_{cl}}{W_L}\right) + |M|_{wL}$$

- In a semi log graph sheet, mark frequency on X-axis and Magnitude in Y-axis. Mark all the points and join them.
- Phase Plot: Calculate the various phase values of w and plot the values in the same graph and join the points by smooth curve.

Calculate the Gain Margin and Phase Margin from the Graph,

$$\text{Gain Margin} = \frac{1}{20\log|G(jw)|_{wpc}} [\text{or}] - 20\log|G(jw)|_{wpc}$$

Where, w_{pc} = phase cross over frequency (i.e frequency at which the phase crosses -180°).

$$\text{Phase Mar gin} = 180 + \varphi_{gc}$$

Where,

$$\varphi_{gc} = \text{phase angle at } w_{gc}\,.$$

(i.e., the angle at which the magnitude crosses 0dB [or] gain cross over frequency w_{gc}).

Advantages of Bode plot:

- Magnitudes are expressed in dB and so, a simple procedure is available to add the magnitude of each term one by one.
- Frequency domain specifications can be easily determined.
- Used to analyze both the open loop and closed loop system.

Problems

Model 1: Bode Problem

1. Let us draw the bode plot for the system GH(s) = 1/s (1+0.2s)(1+0.02s). Also we shall analyze the gain margin and phase margin.

Solution:

Given:

$$GH(s) = 1/s(1+0.2s)(1+0.02s)$$

$$|M|_{wL} = 20\log\left|\frac{1}{jw}\right|:$$

$$|M|_{wCl} = \text{Change in slope from } W_{cl} \text{ to } W_L * \log\left(\frac{W_{cl}}{W_L}\right) + |M|_{wL}$$

$$|M|_{wC2} = \text{Change in slope from } W_{c2} \text{ to } W_{c1} * \log\left(\frac{W_{c2}}{W_{cl}}\right) + |M|_{W_{cl}}$$

$$|M|_{wH} = \text{Change in slope from } W_H \text{ to } W_{c2} * \log\left(\frac{W_H}{W_{c2}}\right) + |M|_{W_{c4}}$$

Step (i): Convert the GH(s) into bode form and substitute s = jw,

$$GH(jw) = \frac{1}{jw(1+j0.2)(1+j0.02w)}$$

Step (ii): The corner frequencies are,

W_{c1} = 1/0.2 = 5 rad/sec

W_{c2} = 1/0.02 = 50 rad/sec.

Choose the lowest and highest corner frequencies,

W_L = 0.1 rad/sec

W_H = 100 rad/sec

Step (iii): Magnitude Plot.

Term	Corner Frequency	Slope	Change in Slope
1/jw	--	-20dB/dec	
1/(1+j0.2w)	5 rad/sec	-20 dB/dec	-20-20 = -40dB/dec
1/(1+j0.02w)	50 rad/sec	-20 dB/dec	-40-20 = -60dB/dec

Step (iv) Obtain magnitude calculation for each corner frequencies:

Summary to Plot Magnitude	
Corner Freq	Magnitude
W_L = 0.1	20 dB
W_{c1}=5	-14 dB

$W_{c2} = 50$	-54 dB
$W_H = 100$	-72 dB

$$|M|_{wL} = 20\log\left|\frac{1}{jw}\right| = 20\log\left|\frac{1}{0.1}\right| = 20\text{ dB}$$

$$|M|_{wCl} = \text{Change in slope from } W_{cl} \text{ to } W_L * \log\left(\frac{W_{cl}}{W_L}\right) + |M|_{wL}$$

$$= -20\text{dB} * \log\left(\frac{5}{0.1}\right) + 20\text{dB}$$

$$= -14\text{ .dB}$$

$$|M|_{wC2} = \text{Change in slope from } W_{c2} \text{ to } W_{c1} * \log\left(\frac{W_{c2}}{W_{cl}}\right) + |M|_{W_{cl}}$$

$$= -40\text{dB} * \log\left(\frac{50}{5}\right) - 14\text{dB}$$

$$= -54\text{ dB}$$

$$|M|_{wH} = \text{Change in slope from } W_H \text{ to } W_{c2} * \log\left(\frac{W_H}{W_{c2}}\right) + |M|_{W_{c4}}$$

$$= -60\text{dB} * \log\left(\frac{100}{50}\right) - 54\text{dB}$$

$$= -72\text{ dB}$$

Step (v): To obtain the Bode phase plot,

$$\angle G(jw) = -90 - \tan^{-1}(0.2w) - \tan^{-1}(0.02w)$$

w	0.1	1	5	10	50	100
$\angle G(jw)$	-91.265	-102.45	-140.7	-164.7	-219.3	-240.5
Mag G(jw)	20dB		-14dB		-54dB	-72dB

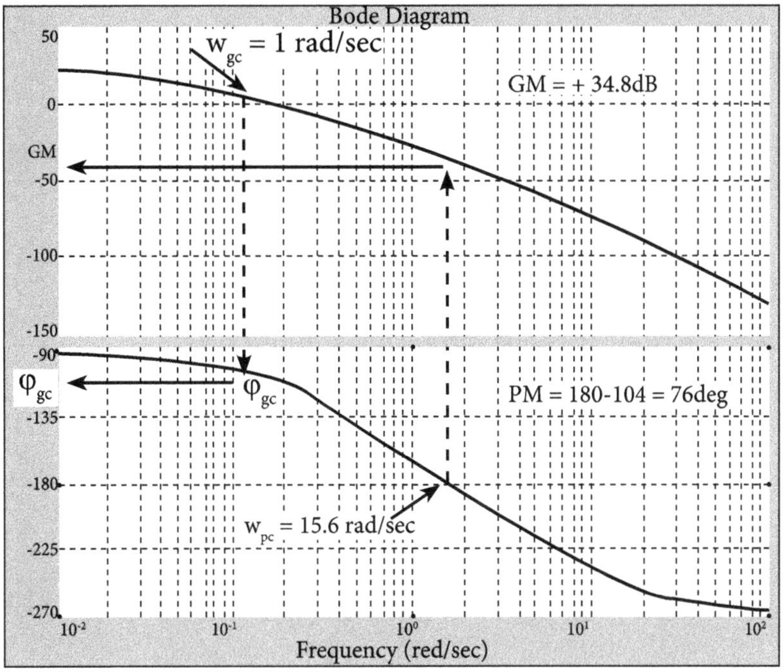

From the Bode plot, w_{gc} = 1 rad/sec; w_{pc} = 15.6 rad/sec.

Gain Margin = +34.8dB & Phase Margin = 180 + φ_{gc} = 180 – 104 = 76°.

Model 2: Bode Problem

2. Let u s d raw the bode plot for the system $G(s)=\frac{K}{s(s+4)(s+10)}$ with the phase margin 30°.

Solution:

Given:

$$G(s)=\frac{K}{s(s+4)(s+10)}$$

Phase margin = 30°

Formula to be used:

$$|M|_{wL}=20\log\left|\frac{1}{jw}\right|$$

$$|M|_{wCl}=\text{Change in slope from } W_{cl} \text{ to } W_L * \log\left(\frac{W_{cl}}{W_L}\right)+|M|_{wL}$$

$$|M|_{wC2} = \text{Change in slope from } W_{c2} \text{ to } W_{c1} * \log\left(\frac{W_{c2}}{W_{cl}}\right) + |M|_{W_{cl}}$$

$$|M|_{wH} = \text{Change in slope from } W_{H} \text{ to } W_{c2} * \log\left(\frac{W_{H}}{W_{c2}}\right) + |M|_{W_{c4}}$$

$PM = 180° + \varphi_{gc}$

Step (i): Convert the G(s) into bode form & substitute s = jw,

$$G(s) = \frac{K}{s4x\left(1+\frac{s}{4}\right)10\left(1+\frac{s}{10}\right)} = \frac{K}{40*s\left(1+\frac{s}{4}\right)\left(1+\frac{s}{10}\right)}$$

$$G(jw) = \frac{K}{40*jw(1+j0.25w)(1+j0.1w)} = \frac{K}{40}*\frac{1}{jw(1+j0.25w)(1+j0.1w)}$$

Let assume K/40 =1, then solve the problem.

Step (ii): The corner frequencies are,

$$W_{c1} = 1/0.25 = 4\,\text{rad/sec}$$

$$W_{c2} = 1/0.1 = 10\ \text{rad/sec}$$

Choose the lowest and highest corner frequencies.

$$W_L = 0.1\ \text{rad/sec}$$

$$W_H = 20\ \text{rad/sec}$$

Step (iii): Magnitude Plot,

Term	Corner Frequency	Slope	Change in Slope
K/40(jw)	--	-20dB/dec	
1/(1+j0.25w)	4 rad/sec	-20 dB/dec	-20-20 = -40dB/dec
1/(1+j0.1w)	10 rad/sec	-20 dB/dec	-40-20 = -60dB/dec

Summary to Plot Magnitude	
Corner Freq	Magnitude
$W_L = 0.1$	20 dB
$W_{c1}=4$	-12 dB

$W_{c2} = 10$	-28 dB
$W_H = 20$	-52 dB

Step (iv): Obtain magnitude calculation for each corner frequencies,

$$|M|_{wL} = 20\log\left|\frac{1}{jw}\right| = 20\log\left|\frac{1}{0.1}\right| = 20\text{ dB}$$

$$|M|_{wCl} = \text{Change in slope from } W_{cl} \text{ to } W_L * \log\left(\frac{W_{cl}}{W_L}\right) + |M|_{wL}$$

$$= -20\text{dB} * \log\left(\frac{4}{0.1}\right) + 20\text{dB}$$

$$= -12.04\text{ dB}$$

$$|M|_{wC2} = \text{Change in slope from } W_{c2} \text{ to } W_{c1} * \log\left(\frac{W_{c2}}{W_{cl}}\right) + |M|_{W_{cl}}$$

$$= -40\text{dB} * \log\left(\frac{10}{4}\right) - 12\text{dB}$$

$$= -28\text{ dB}$$

$$|M|_{wH} = \text{Change in slope from } W_H \text{ to } W_{c2} * \log\left(\frac{W_H}{W_{c2}}\right) + |M|_{W_{c4}}$$

$$= -60\text{dB} * \log\left(\frac{20}{10}\right) - 34\text{dB}$$

$$= -52\text{ dB}$$

Step (v): To obtain the Bode phase plot,

$$\angle G(jw) = -90 - \tan^{-1}(0.25w) - \tan^{-1}(0.1w)$$

W	0.1	1	2	4	6	10	20
$\angle G(jw)$	-92°	-110°	-128°	-156°	-177°	-203°	-232°
	20dB			-12dB		-28dB	-52dB

Given phase margin = 30°.

We know that,

$$PM = 180^\circ + \varphi_{gc}$$

$$\Phi_{gc} = 30^\circ - 180^\circ = -150^\circ.$$

From the bode plot, the magnitude corresponds to the angle –150° is –12 dB.

Therefore,

20log (K/40) = +12 dB.

K = 40*Anti log (12/20) = 40*3.98.

K = 159.24 is the required value for phase margin of 30°.

3. Let us calculate the value of k for open loop transfer function of a unity feedback system $G(s)=\frac{K^{e-0.1s}}{s(1+0.1s)(1+s)}$ so that the phase margin of the system is 60°.

Solution:

Given:

$$G(s)=\frac{Ke^{-0.1s}}{s(1+0.1s)(1+s)}$$

Phase margin = 60°

Formula to be used:

$$|M|_{wL}=20\log\left|\frac{1}{jw}\right|$$

$$|M|_{wCl}=\text{Change in slope from } W_{cl} \text{ to } W_L * \log\left(\frac{W_{cl}}{W_L}\right)+|M|_{wL}$$

$$|M|_{wC2}=\text{Change in slope from } W_{c2} \text{ to } W_{c1} * \log\left(\frac{W_{c2}}{W_{cl}}\right)+|M|_{W_{cl}}$$

$$|M|_{wH}=\text{Change in slope from } W_H \text{ to } W_{c2} * \log\left(\frac{W_H}{W_{c2}}\right)+|M|_{W_{c4}}$$

$$PM = 180^\circ + \phi_{gc}$$

(i) Substitute s = jw, let assume K = 1,

$$GH(jw)=\frac{\left(e^{-0.1jw}\right)}{jw(1+j0.1w)(1+jw)}$$

(ii) The corner frequencies are,

$$W_{c1}=1/1=1\text{ rad/sec}$$

$$W_{c2}=1/0.1=10\text{ rad / sec}$$

(iii) Choose the lowest and highest corner frequencies,

$$W_L=0.1\text{ rad / sec}$$

$$W_H=20\text{ rad / sec.}$$

Magnitude Plot

Term	Corner Frequency	Slope	Change in Slope
1/jw	--	-20dB/dec	
1/(1+jw)	1 rad/sec	-20 dB/dec	-20-20 = -40dB/dec
1/(1+j0.1w)	10 rad/sec	-20 dB/dec	-40-20 = -60dB/dec

Summary to Plot Magnitude	
Corner Freq	Magnitude
$W_L = 0.1$	20 dB
$W_{c1}=1$	0 dB
$W_{c2}= 10$	-40 dB
$W_H = 20$	-58 dB

(iv) Obtain Magnitude Calculation for each corner frequencies,

$$|M|_{wL}=20\log\left|\frac{1}{jw}\right|=20\log\left|\frac{1}{0.1}\right|=20\text{ dB}$$

$$|M|_{wCl}=\text{Change in slope from } W_{cl}\text{ to } W_L * \log\left(\frac{W_{cl}}{W_L}\right)+|M|_{wL}$$

$$=-20\text{dB}*\log\left(\frac{1}{0.1}\right)+20\text{dB}$$

$$=0\text{ dB}$$

$$|M|_{wC2} = \text{Change in slope from } W_{c2} \text{ to } W_{c1} * \log\left(\frac{W_{c2}}{W_{cl}}\right) + |M|_{W_{cl}}$$

$$= -40\,dB * \log\left(\frac{10}{1}\right) + 0\ dB$$

$$= -40\ dB$$

$$|M|_{wH} = \text{Change in slope from } W_{H} \text{ to } W_{c2} * \log\left(\frac{W_{H}}{W_{c2}}\right) + |M|_{W_{c4}}$$

$$= -60\,dB * \log\left(\frac{20}{10}\right) - 30\,dB$$

$$= -58\ dB$$

(v). To obtain the Bode phase plot,

$$\angle G(jw) = -0.1w\frac{180}{\pi} - 90 - \tan^{-1}(0.5w) - \tan^{-1}(0.125w)$$

$$\angle G(jw) = -5.729w - 90 - \tan^{-1}(0.5w) - \tan^{-1}(0.125w)$$

W	0.1	1	2	5
$\angle G(jw)$	-96.8°	-146.4°	-176.2°	-223.8°

Given PM = 60°,

$$P.M = 180° + \phi_{gc}$$

$$60° = 180° + \phi_{gc} \text{ gives } \phi_{gc} = -120°$$

From the bode plot, the magnitude value |G(jwpc)| corresponds to ϕ_{gc} is -8dB

Therefore,

20 log K = - 8dB K = Anti log (-8/20)

$$K = 10^{\left(\frac{-8}{20}\right)} = 0.398 \approx 0.4$$

Therefore, the value of K for PM = 60° is 0.4.

4. Let us draw the bode plot with the gain margin 6dB and the p ha se margin 45° for the given open loop transfer function of a unity feedback system ,

$$G(s)=\frac{Ke^{-0.2s}}{s(s+2)(s+8)}$$

Solution:

Given:

$$G(s)=\frac{Ke^{-0.2s}}{s(s+2)(s+8)}$$

Phase margin = 45°

Formula to be used:

$$|M|_{wL}=20\log\left|\frac{1}{jw}\right|$$

$$|M|_{wCl}=\text{Change in slope from } W_{cl} \text{ to } W_L * \log\left(\frac{W_{cl}}{W_L}\right)+|M|_{wL}$$

$$|M|_{wC2}=\text{Change in slope from } W_{c2} \text{ to } W_{c1} * \log\left(\frac{W_{c2}}{W_{cl}}\right)+|M|_{W_{cl}}$$

$$|M|_{wH}=\text{Change in slope from } W_H \text{ to } W_{c2} * \log\left(\frac{W_H}{W_{c2}}\right)+|M|_{W_{c4}}$$

$$PM=180°+\phi_{gc}$$

(i) Substitute s = jw,

$$GH(jw)=\frac{Ke^{-0.2jw}}{2\times 8jw\left(1+\frac{jw}{2}\right)\left(1+\frac{jw}{8}\right)}$$

$$GH(jw)=\frac{K}{16}*\frac{e^{-0.2jw}}{jw(1+j0.5w)(1+0.125w)}$$

(ii) The corner frequencies are,

$$W_{c1}=1/0.5s=2 \text{ rad/sec}$$

$$W_{c2} = 1/0.125 = 8 \text{ rad/sec}$$

(iii) Choose the lowest and highest corner frequencies,

$$W_L = 0.1 \text{ rad/sec}$$

$$W_H = 20 \text{ rad/sec}$$

Magnitude Plot

Term	Corner Frequency	Slope	Change in Slope
1/jw	--	-20dB/dec	
1/(1+j0.5w)	2 rad/sec	-20 dB/dec	-20 -20 = -40dB/dec
1/(1+j0.125w)	8 rad/sec	-20 dB/dec	-40-20 = -60dB/dec

3. Obtain Magnitude Calculation for each corner frequencies,

$$|M|_{wL} = 20\log\left|\frac{1}{jw}\right| = 20\log\left|\frac{1}{0.1}\right| = 20 \text{ dB}$$

$$|M|_{wCl} = \text{Change in slope from } W_{cl} \text{ to } W_L * \log\left(\frac{W_{cl}}{W_L}\right) + |M|_{wL}$$

$$= -20\text{dB} * \log\left(\frac{2}{0.1}\right) + 20\text{dB}$$

$$= -6.02 \text{ dB}$$

$$|M|_{wC2} = \text{Change in slope from } W_{c2} \text{ to } W_{c1} * \log\left(\frac{W_{c2}}{W_{cl}}\right) + |M|_{W_{cl}}$$

$$= -40\text{dB} * \log\left(\frac{8}{2}\right) + -6.02 \text{ dB}$$

$$= -30.10 \text{ dB}$$

$$|M|_{wH} = \text{Change in slope from } W_H \text{ to } W_{c2} * \log\left(\frac{W_H}{W_{c2}}\right) + |M|_{W_{c4}}$$

$$= -60\text{dB} * \log\left(\frac{20}{8}\right) - 30\text{dB}$$

$$= -53.97 \text{ dB}$$

4. To obtain the Bode phase plot,

$$\angle G(jw) = -0.2w\frac{180}{\pi} - 90 - \tan^{-1}(0.5w) - \tan^{-1}(0.125w)$$

$$\angle G(jw) = -11.45w - 90 - \tan^{-1}(0.5w) - \tan^{-1}(0.125w)$$

W	0.1	1	2	5	8	10	20
$\angle G(jw)$	-94.72°	-135.14°	-171.9°	-247.4°	-302.5°	-334.5°	-471.4°
Mag	20dB		-6dB		-30dB		-54dB

From the bode plot, W_{gc} = 1rad/sec, W_{pc} = 2rad/sec, GM = 34 dB.

Given,

PM = 45°, but we know that PM = 180° + φ_{gc} = 45° - 180° = -135°

From the bode plot, the magnitude value |G (jw)| corresponds to φ_{gc} is 0dB.

Therefore,

20 log (K/16) = 0

K = 16*Anti-log (0/20) = 16

Given Gain margin = 6 dB,

But actual gain margin from the bode plot = 10 dB.

20 log K = GM actual – GM given 20 log (K/16) = 10 - 6

Therefore,

20log (K/16) = 4

K = 16*Anti log (4/2 0) = 25.12.

Model 3: Bode Problem

5. Let us calculate the value of K with Gain Margin = 20 dB and Phase margin = 60° for the open loop transfer function of a unity feedback system $G(s) = \frac{K}{s(s+2)(s+4)}$.

Solution:

Given:

$$G(s) = \frac{K}{s(s+2)(s+4)}$$

Gain Margin = 20 dB

Phase margin = 60°

Formula to be used:

$$PM = 180° + \varphi_{gc}$$

$$G(s)=\frac{K}{s(s+2)(s+4)}$$

Substitute s = jw, then obtain the magnitude and phase expression.

$$G(jw)=\frac{K}{jw(jw+2)(jw+4)}$$

$$\text{Mganitude}\left|G(jw)\right|=\frac{K}{w\sqrt{w^2+4}\sqrt{w^2+16}};$$

$$\text{Phase Angle } \angle G(jw)=-90°-\tan^{-1}\left(\frac{w}{2}\right)-\tan^{-1}\left(\frac{w}{4}\right)$$

Given Gain Margin = 20 dB or when the gain margin is given, the phase angle is 180°, therefore,

$$\angle G(jw)=-90°-\tan^{-1}\left(\frac{w}{2}\right)-\tan^{-1}\left(\frac{w}{4}\right)=180°$$

Hence,

$$\tan^{-1}\left(\frac{w}{2}\right)+\tan^{-1}\left(\frac{w}{4}\right)=90°$$

Take tan on both sides,

$$\tan\left[\tan^{-1}\left(\frac{w}{2}\right)+\tan^{-1}\left(\frac{w}{4}\right)\right]=\tan 90°$$

$$\left(\frac{\frac{w}{2}+\frac{w}{4}}{1-\frac{w}{2}\times\frac{w}{4}}\right)=\tan 90°$$

$$\frac{\frac{w}{2}+\frac{w}{4}}{1-\frac{w^2}{8}}=\tan 90°$$

$$\frac{\frac{w}{2}+\frac{w}{4}}{1-\frac{w^2}{8}}=\infty \Rightarrow 1-\frac{w^2}{8}=0, \Rightarrow w_{pc}=\sqrt{8}$$

The magnitude at phase cross over frequency $W_{pc} = \sqrt{8}$ is given by,

$$\left|G(jw)\right|_{w=w_{pc}}=\frac{K}{w_{pc}\sqrt{w_{pc}^2+4}\sqrt{w_{pc}^2+16}}=\frac{K}{\sqrt{8\times(12)\times(24)}}=\frac{K}{48}$$

For the given GM, the magnitude at W_{pc} = 0.1, then the value of K can be obtained as,

$$\frac{K}{48}=0.1 \Rightarrow K=4.8$$

Given Phase Margin = 60°.

We know that P.M = 180° + φ_{gc} = φ_{gc} = 60°-180° = -120°.

The gain cross over frequency, W_{gc} at this φ_{gc} can be obtained as,

$$\text{P.M}=180°+\angle G(jw) \text{ or } \varphi_{gc}$$

$$-120°=-90°-\tan^{-1}\left(\frac{w_{gc}}{2}\right)-\tan^{-1}\left(\frac{w_{gc}}{4}\right)$$

$$\tan^{-1}\left(\frac{w_{gc}}{2}\right)+\tan^{-1}\left(\frac{w_{gc}}{4}\right)=30°$$

Take tan on both sides,

$$\tan\left[\tan^{-1}\left(\frac{w_{gc}}{2}\right)+\tan^{-1}\left(\frac{w_{gc}}{4}\right)\right]=\tan 30°$$

$$\frac{\frac{w_{gc}}{2}-\frac{w_{gc}}{4}}{1+\frac{w_{gc}}{2}\times\frac{w_{gc}}{4}}=\tan 30°=0.577$$

$$\frac{6w_{gc}}{8-w_{gc}^2}=0.577 \Rightarrow 0.577\,w_{gc}^2+6w_{gc}=8\times 0.577$$

$W_{gc}^2+10.4w_{gc}=8$, by solving this quadratic equation we get $w_{gc}=0.719$

At gain cross over frequency, the $\left|G(jw_{gc})\right|=1$,

Hence,

$$\left|G(jw_{gc})\right|=\frac{K}{w_{gc}\sqrt{w_{gc}^2+4}\sqrt{w_{gc}^2+16}}=1$$

$$=\frac{K}{0.791\sqrt{(0.719^2+4)(0.719^2+16)}}=1 \Rightarrow K=6.21$$

The value of K for the Gain Margin = 20 dB is K = 4.8.

The value of K for the Phase Margin = 60° is K = 6.21.

6. Let us sketch the bode plot for the following transfer function $G(s)=\frac{K(s+3)}{s(s+1)(s+2)}$.

Solution:

Given:

$$G(s)=\frac{K(s+3)}{s(s+1)(s+2)}$$

Formula to be used:

$$A=20\log\left|\frac{1.5}{j\omega}\right|$$

Magnitude Plot

$$G(s)=\frac{K\cdot 3\,(S/3+1)}{S(1+s)2(3/2+1)} \Rightarrow \frac{3K(1+0.33s)}{23(1+3)(1+0.5s)}$$

$$=\frac{1.5K(1+0.33s)}{S(1+S)(1+0.5s)}.$$

Let,

$$K=1, \quad G(s)=\frac{1.5(1\ \ 0.33s)}{S(1+S)(1+0.5s)}$$

Put,

$$s=j\omega; \quad G(j\omega)=\frac{1.5(1+0.33j\omega)}{j\omega(1+j\omega)(1+0.5j\omega)}$$

The corner frequencies are,

$$\omega_{cl}=\frac{1}{1}=1 \text{ rad/sec}; \quad \omega_{c2}=\frac{1}{0.5}=2 \text{ rad/sec}.$$

$$\omega_{c3}=\frac{1}{0.33}=3 \text{ rad/sec}.$$

Term	Corner Frequency rad/sec	Slope dB/dec	Change in slope dB/dec
$1.5/j\omega$	–	−20	–
$\frac{1}{(1+j\omega)}$	1	−20	−40
$\frac{1}{(1+0.5\omega)}$	2	−20	−60
$(1+0.33j\omega)$	3	20	−40

Let low frequency $\omega_l = 0.1$ rad / sec .

Let high frequency, $\omega_h = 50$ rad / sec.

Let us calculate the magnitude for the above frequency.

$$\text{At } \omega=\omega_l, A=20\log\left|\frac{1.5}{j\omega}\right|=20\log\left|\frac{1.5}{0.1}\right|=23.52 \text{ db}$$

$$\text{At } \omega=\omega_d \Rightarrow A=-40\times\log\left(\frac{2}{1}\right)+3.52=-8.52 \text{db}$$

$$\text{At } \omega=\omega_{c3} \Rightarrow A=-60\times\log\left(\frac{3}{2}\right)-8.52=-19 \text{db}$$

$$\text{At } \omega=\omega_h \Rightarrow A=-40\times\log\left(\frac{50}{3}\right)-19=-68 \text{db}.$$

Phase Plot

$$\phi = \tan^{-1}(0.33\,\omega) - 90° - \tan^{-1}(\omega) - \tan^{-1}(0.5\omega)$$

ω	φ
0.1	−96°
1	−143°
2	−165°
3	−173°
50	−180°

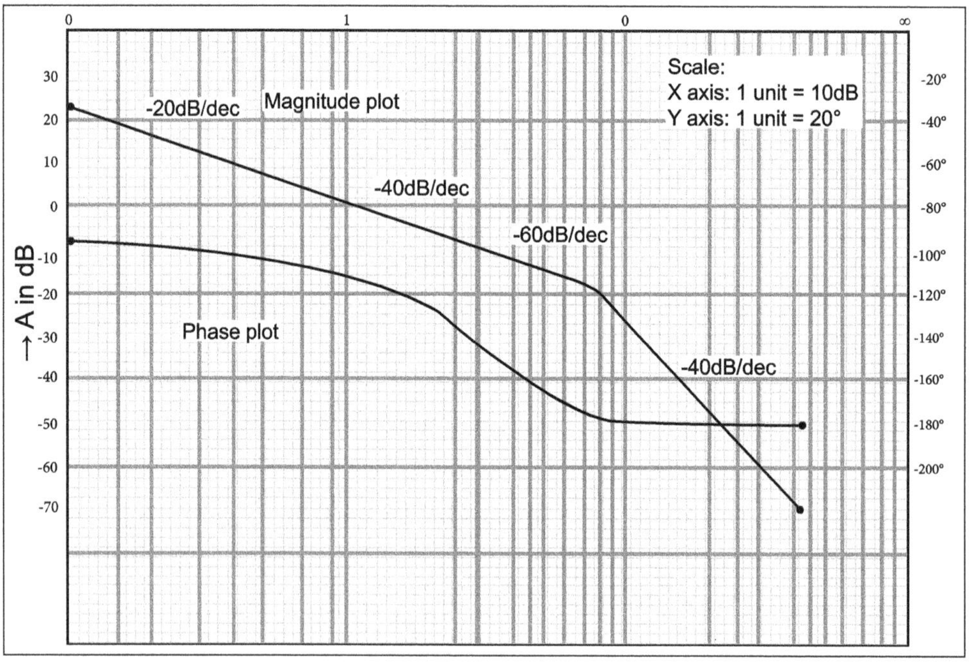

4.3 Polar Plots

Polar plots can be used to predict the feedback control system stability by the application of NY Quist Criterion and therefore, it can also be referred as NY Quist Plots. It is a labor saving technique in the analysis of dynamic behavior of control systems in which the need for finding the roots of characteristic equation of the system is eliminated. Let s draw the polar plot of an integral term transfer function.

Integral term transfer function, G(s) = 1/s

$$\therefore \; G(j\omega) = 1/j\omega$$

$$\therefore \; \text{Magnitude} = 1/\omega$$

Angle, $\varphi = -90^\circ$

Therefore, the locus is on the negative frequency axis.

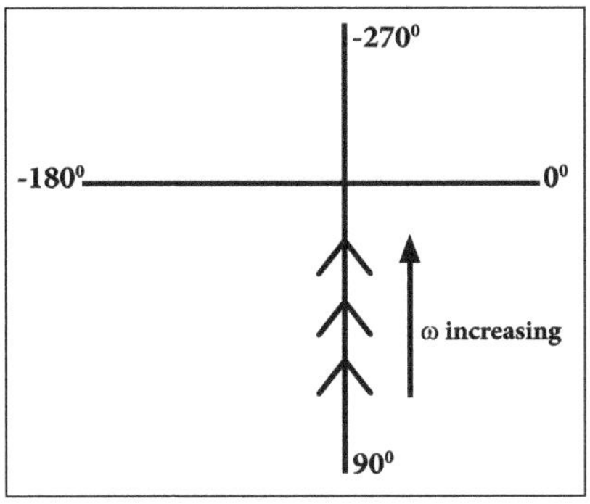

Locus plot.

Problems

1. Let us draw the polar plot of quadratic equation $G(j\omega)=\dfrac{1}{1+2\xi\left(\dfrac{j\omega}{\omega_n}\right)+\left(\dfrac{j\omega}{\omega_n}\right)^2}$; for $\xi > 0$.

Solution:

Given:

$$G(j\omega)=\frac{1}{1+2\xi\left(\frac{j\omega}{\omega_n}\right)+\left(\frac{j\omega}{\omega_n}\right)^2};\ \text{for } \xi > 0.$$

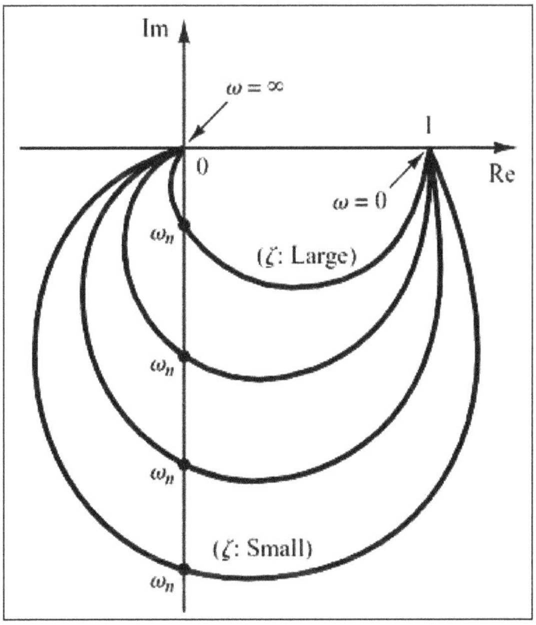

2. Let us draw the polar plot to calculate the frequency at $|G(j\omega)| = 1$. If open loop TF of unity feedback system is given by $G(s)=\frac{1}{s(s+1)^2}$.

Solution:

Given:

$$G(s)=\frac{1}{s(s+1)^2};\ H(s)=1.$$

$$|G(j\omega)|=1$$

Polar Plot

$$G(s)=\frac{1}{s(s+1)^2};H(s)=1.$$

Put,

$$s=j\omega$$

$$\therefore\ G(j\omega)=\frac{1}{j\omega(j\omega+1)^2}=\frac{1}{j\omega(1+j\omega)(1+j\omega)}.$$

Convert frequency,

$$\omega_{c1}=1\ \text{rad/sec.}$$

$$G(j\omega)=\frac{1}{j\omega(1+j\omega)^2}=\frac{1}{j\omega(1+j\omega)(1+j\omega)}.$$

$$G(j\omega)=\frac{1}{\omega\angle 90^\circ\sqrt{1+\omega^2}\ \angle\tan^{-1}\omega\sqrt{1+\omega^2}\ \angle\tan^{-1}\omega}$$

$$|G(j\omega)|=\frac{1}{\omega\left(1+\omega^2\right)}=\frac{1}{\omega+\omega^3}$$

$$\angle G(j\omega)=-90^\circ-2\tan^{-1}\omega$$

ω rad/sec	0.4	0.5	0.6	0.7	0.8	0.9	1.0	1.1

$\lvert G(j\omega)\rvert$	2.2	1.6	1.2	1	0.8	0.6	0.5	0.4
$\angle G(j\omega)$ degree	−134	−143	−151	−159	−167	−174	−180	−185

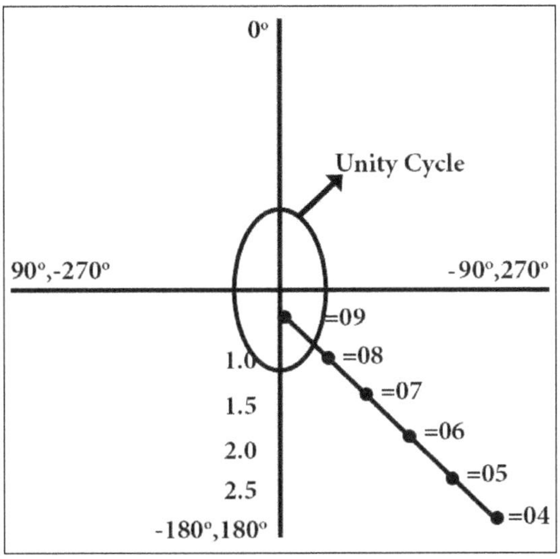

$$\text{Gain margin, kg} = \frac{1}{\lvert G(j\omega_{pc})\rvert} = \frac{1}{0.5} = 2.$$

4.3.1 Mathematical Preliminaries

The mathematical preliminaries and techniques necessary for generating polar plots and NY Quist stability plots of feedback control systems and the mathematical basis and properties of the NY Quist stability criterion.

Plotting Complex Functions of a Complex Variable

A real function of a real variable is easily graphed on a single set of co - ordinate axes. For example, the real function f(x) is easily plotted in the rectangular co - ordinates with x as the abscissa and f(x) as the ordinate. A complex function of a complex variable, such as the transfer function P(s) with $s = \sigma + j\,\omega$ cannot be plotted on the single set of co - ordinates.

The complex variable $s = \sigma + j\,\omega$ depends on two independent quantities, the real and imaginary parts of s. Hence s cannot be represented by a line. The complex function P(s) also has real and imaginary parts. Similarly, the complex variable $z = \mu + jv$ and discrete-time system complex transfer functions P (z) cannot be graphed in one dimension.

In general, in order to plot P(s) with $s = \sigma + j\,\omega$, two two-dimensional graphs are required. The first is the graph of jω versus σ called the s-plane, the same set of co - ordinates as those used for plotting the pole-zero maps. The second is the imaginary part of P(s) (I m P) versus the real part of P(s) (Re P) termed as P(s)-plane. The corresponding co - ordinate planes f or discrete-time systems are the z- plane and the P(z)-plane.

The correspondence between points in the two planes is called a s mapping or transformation. For example, points in the s-plane are mapped into points of the P(s)-plane by the function P.

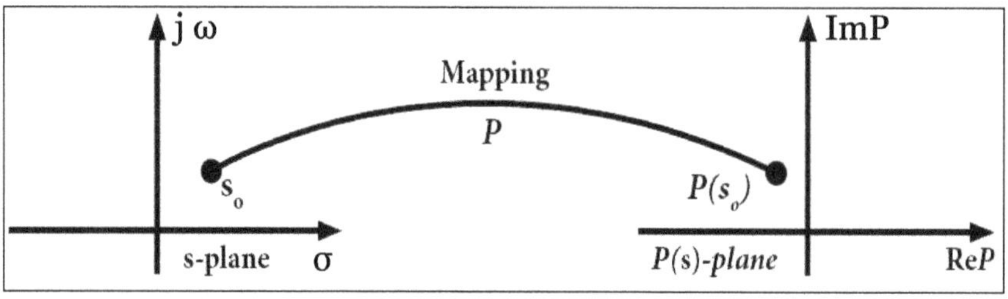

Mapping

In general, only a very specific locus of points in the s-plane is mapped into the P(s) - plane. For NY Quist stability plots, this locus is called NY Quist path. For the special case σ = 0, s = j ω , the s-plane degenerates into a line and P(jω) may be represented in a P(jω)-plane with ω as a parameter. Polar p lots are constructed in the P(jω)-plane from this line (s = jω) in the s-plane.

4.4 NY Quist Stability Criterion

We know that q(s) = 1+G(s) H(s). Therefore, G(s)H(s) = [1+G(s)H(s)] – 1. The contour which has obtained due to mapping of NY Quist contour from s-plane to q(s)-plane. (i.e.,) [1+G(s) H(s)] plane will encircle about the origin.

The contour CGH, which has obtained due to mapping of NY Quist contour from s-plane to G(s)H(s)-plane, will encircle about the point (-1+j0). Therefore, encircling the origin in the q(s)-plane is equivalent to encircling the point -1+j0 in the G(s) H(s)-plane.

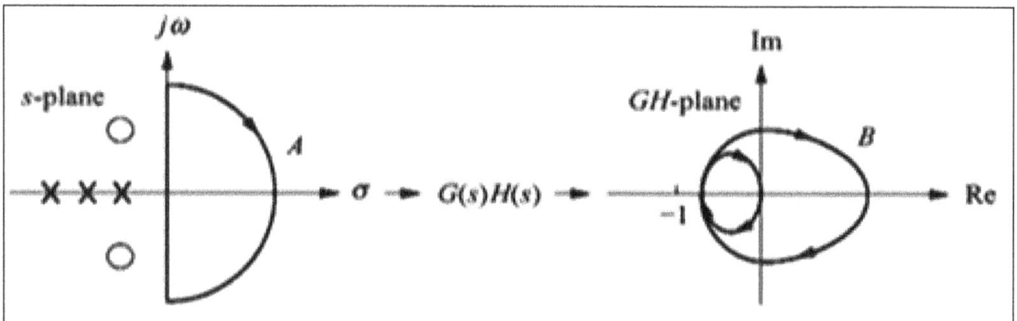

(a)

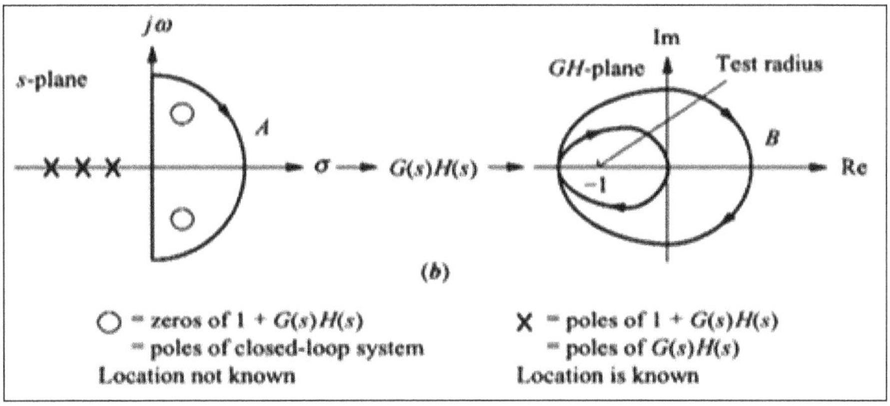

NY Quist stability criterion says that N = Z - P where N is the total n umber of encirclement about the origin, P is the total n umber of poles and Z is the total n umber of zeros.

- Case 1: N = 0 (no encirclement), so Z = P = 0 & Z = P. If N = 0, P must be zero therefore system is stable.
- Case 2: N > 0 (clockwise encirclement), so P = 0, Z ≠ 0 & Z > P. For both the cases, system is unstable.
- Case 3: N < 0 (counterclockwise encirclement), so Z = 0, P ≠ 0 & P > Z. System is stable.

Advantages of NY Quist stability criterion over Routh's Hurwitz criteria:

- Nyquist stability criterion allows us to deal with any analytical function directly.
- Nyquist stability criterion provides a measure of the degree of stability of the system and shows the frequency range of the system.

Problems

1. Let us sketch the Nyquist stability plot for a feedback system with the following open-loop transfer function $G(s)H(s)=\frac{1}{s(s^2+s+1)}$.

Solution:

Given:

$$G(s)H(s)=\frac{1}{s(s^2+s+1)}$$

For section ab, $s = j\omega,\ \omega : 0 \rightarrow \infty$

$$G(j\omega)H(j\omega)=\frac{1}{j\omega\left(1-\omega^2+j\omega\right)}$$

$$\omega\to 0:G(j\omega)H(j\omega)\to -1-j\infty$$

$$\omega=1:G(j\omega)H(j\omega)\to -1-j0$$

$$\omega\to\infty:G(j\omega)H(j\omega)\to 0\angle -270^\circ$$

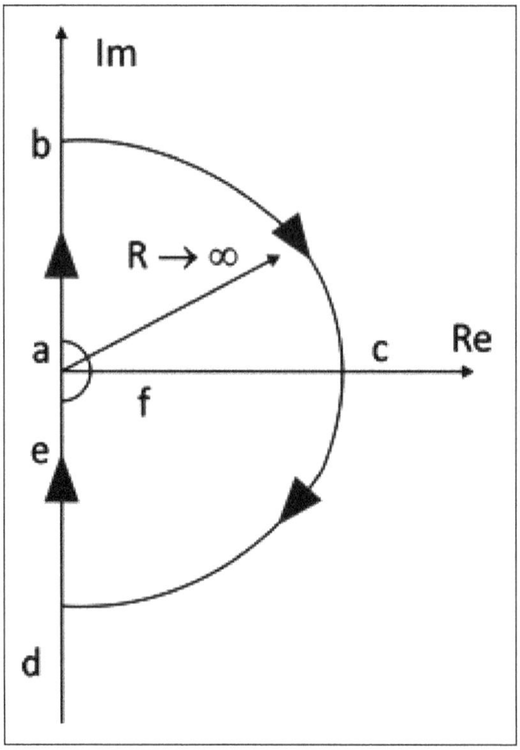

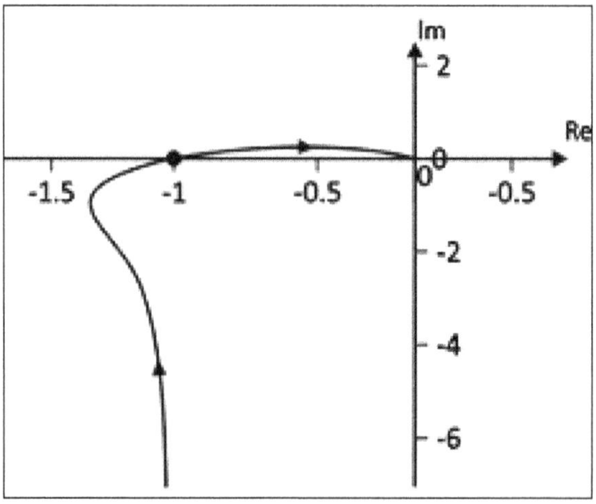

On section b c d, $S = Re^{j\theta}\big|_{R\to\infty}$; therefore i.e. Section b c d maps onto the origin of the G(s)H(s)-plane:

$$|G(s)\,H(s)| \to \frac{1}{R^3} \to 0.$$

Section de maps as the complex image of the polar plot as before:

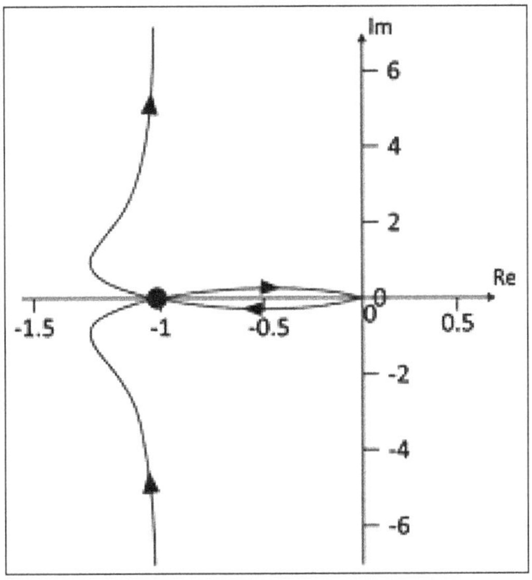

Assessment of relative stability using Nyquist criterion (Systems with transportation lag excluded):

- The stability analysis of a feedback control system is based on identifying the location of the roots of the characteristic equation on the s-plane. The system is stable if the root lies on the left hand side of s-plane. Relative stability of a system can be determined by using the frequency response methods like Nyquist plot and bode plot.
- Nyquist criterion is used to identify the presence of roots of a characteristic equation in a specified region of s-plane.

Nyquist Encirclement: A point is said to be encircled by a contour, if it is found inside the contour.

Nyquist Mapping: The process by which a point in s-plane is transformed into a point in F(s) plane is called mapping and F(s) is called the mapping function.

Steps of Drawing the Nyquist Path

- Step 1 - Check for the poles of G(s) H(s) of jω axis including that at the origin.
- Step 2 – Select the proper Nyquist contour – Include the entire right half of s-plane by drawing a semicircle of radius R with R tends to infinity.

- Step 3 – Identify the various segments on the contour with reference to the Nyquist path.
- Step 4 – Perform the mapping, segment by segment, substituting the equation for respective segment in the mapping functions. Basically we have to sketch the polar plots of the respective segment.
- Step 5 - Mapping of the segments are usually mirror images of mapping of respective path of positive imaginary axis.
- Step 6 - The semicircular path which covers the right half of the s plane generally maps into a point in G(s) H(s) plane.
- Step 7- Interconnect all the mapping of different segments to yield the required Nyquist diagram.
- Step 8 – Note the number of clockwise encirclement about (-1, 0) and decide stability by N = Z – P $G(s)\,H(s) = \frac{N(s)}{D(s)}$ is the open loop transfer function. $\frac{G(s)}{1+G(s)H(s)}$ is the closed loop transfer function.

N(s) = 0 is the open loop zero and D(s) is the open loop pole.

From the stability point of view, no closed loop poles should lie in the RH side of s-plane. Characteristics equation 1 + G(s) H(s) = 0 denotes the closed loop poles.

Let,

$$q(s)=1+G(s)H(s)=\frac{N_1(s)}{D_1(s)}$$

Now as 1+ G(s) H(s) = 0, therefore q(s) should also be zero.

If,

$$q(s)=0,\ \frac{N_1(s)}{D_1(s)}=0 \text{ i,e. } N_1(s)=0$$

Therefore from the stability point of view, zeros of q(s) should not lie in right hand plane of s-plane. To define the stability, entire right hand plane is considered. We assume a semicircle which encloses all points in the right hand plane by considering the radius of the semicircle R tends to infinity. [$R \to \infty$].

The first step to understand the application of Nyquist criterion in relation to the determination of stability of control systems is mapping from s-plane to G(s) H(s) - plane. s is considered as the independent complex variable and corresponding value of G(s) H(s) being a dependent variable plotted in another complex plane called G(s) H(s)-plane. Thus for every point in the s-plane, there exists a corresponding point in G(s) H(s) - plane.

During the process of mapping, the independent variable s is varied along the specified path in the s-plane and the corresponding points in G(s) H(s) plane are joined together. This completes the process of mapping from s-plane to G(s) H(s)-plane.

2. Let us check the stability by Nyquist plot for the following:

$$G_o(s) = \frac{10}{(S+1)(2S+1)(3S+1)}$$

Solution:

Given:

$$G_o(s) = \frac{10}{(S+1)(2S+1)(3S+1)}$$

To find: Stability

Step 1:

First lets draw the Nyquist contour on S - plane enclosing the entire right half of S - plane.

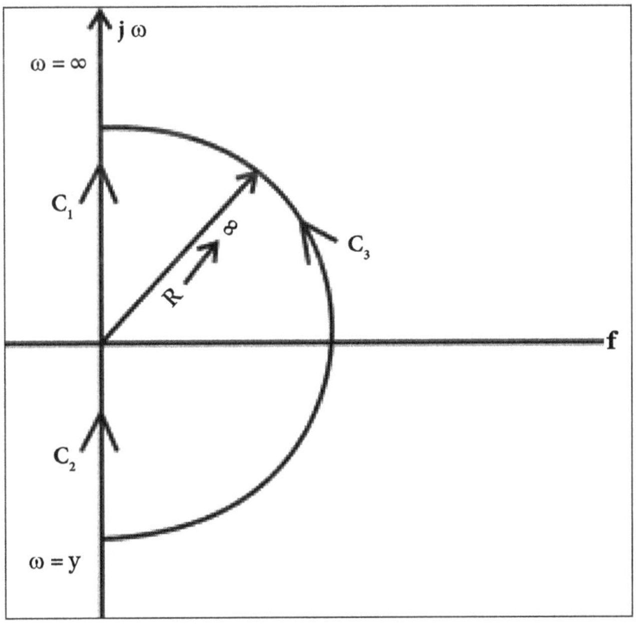

Step 2: Mapping the section C_1,

In section C_1, w varies from 0 to $+\infty$. The mapping of section C_1 is given by the locus of G_0 (jw) as w varies from 0 to ∞. The locus is the polar plot of G_0 (jw).

Type number = 0

Order = 3

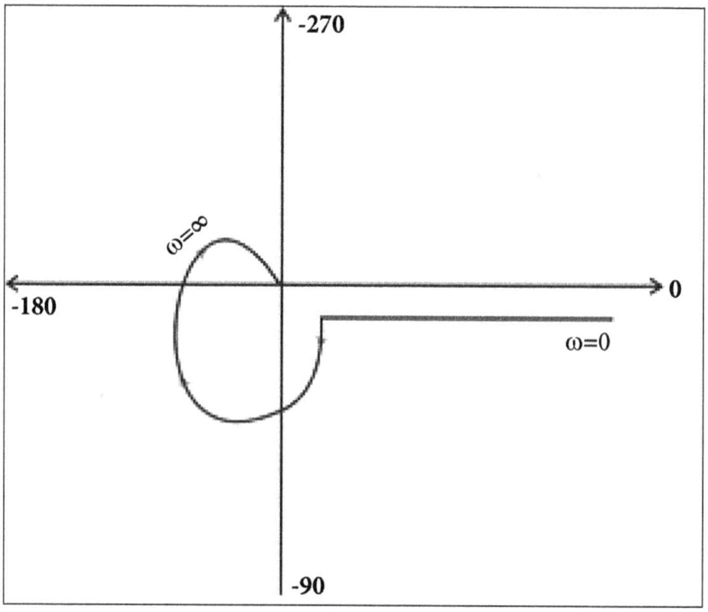

$$G_0(s)=\frac{10}{(1+S)(1+2S)(1+3S)}$$

Put,

$$S=jw$$

$$G_0(jw)=\frac{10}{(1+Sjw)2(jw)^2(1+3jw)}$$

$$=\frac{10}{(1+3jw-2w^2)(1+3jw)}$$

$$=\frac{10}{1+3jw-2w^2+3jw+9(jw)^2-6jw^3}$$

$$=\frac{10}{1+6jw-11w^2-6jw^3}$$

$$=\frac{10}{1-11w^2+6jw(1-w^2)}$$

When the G_0(jw) locus crosses real axis, the imaginary term is zero and the corresponding frequency is the phase cross over frequency.

$$\therefore \text{ At } W = W_{PC}, \quad 6w\left(1-w^2\right)$$

$$1-w^2 = 0$$

$$w^2 = 1$$

$$w = \sqrt{1} = 1 \text{rad/sec}$$

$$\text{At } w = wPC = 1 \text{rad/sec}$$

$$G_0(jw) = \frac{10}{1-11w^2} = \frac{10}{1-11(1)} = \frac{10}{1-11}$$

The G_0 (jw) locus crosses the real axis at a point -1.

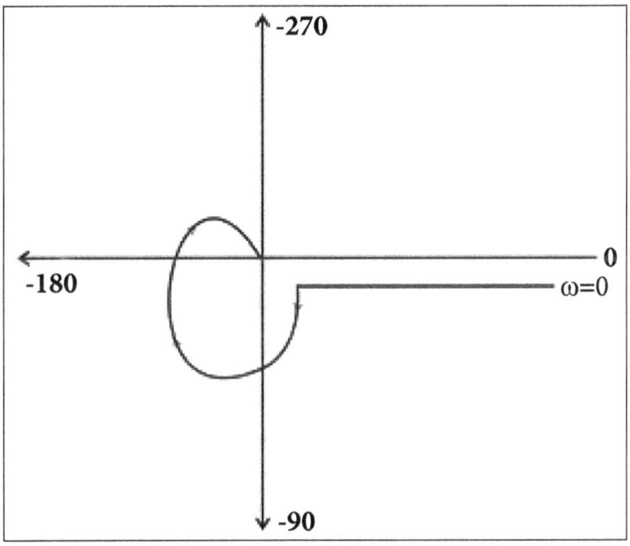

Step 3: Mapping the section C_2,

Let, $S = \lim_{R\to\infty} Re^{jQ}$ in $G_0(S)$ and Varying θ from $+\frac{\pi}{2}$ to $-\frac{\pi}{2}$.

Since $S \to RejQ$ and $R \to \infty G(s)$ can be approximated as show below i.e $(1+sT) = sT$.

$$G_0(s) = \frac{10}{(1+S)(1+2s)(1+3s)} = \frac{10}{S(2S)(3S)} = \frac{10}{6S^3}$$

$$G_o(s)=\frac{10}{6\left(\lim\limits_{R\to\infty} Re^{j\theta}\right)^3}=\frac{10}{\lim\limits_{R\to\infty} 6R^3e^{j3\theta}}$$

$$=\frac{10}{\infty e^{-j3\theta}}=0e^{-j3\theta}$$

$$G_o(s)=0e^{-j3\theta}$$

$$\text{At } \theta=\frac{\pi}{2}\Rightarrow G_o(s)=0e^{-j3\frac{\pi}{2}}$$

$$\text{At } \theta=-\frac{\pi}{2};\ G_o(s)=0e^{-j3\left(\frac{\pi}{2}\right)}=0e^{+j3\left(\frac{\pi}{2}\right)}$$

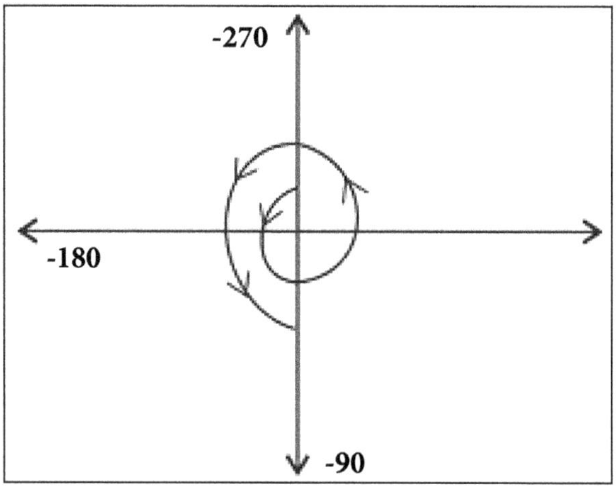

Step 4: Mapping of section C_3,

In section C_3, w varies from -∞ to 0. The mapping of section C_3 is given by the locus of G_o (jw) as w is varied from -∞ to 0. This locus is the inverse polar plot of G_o (jw).

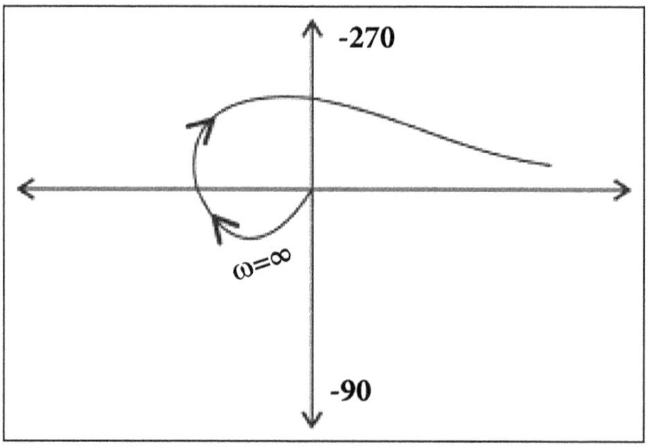

Step 5: Complete Nyquist plot: The entire Nyquist plot in $G_0(S)$ plane can be obtained by combining the mapping of individual section.

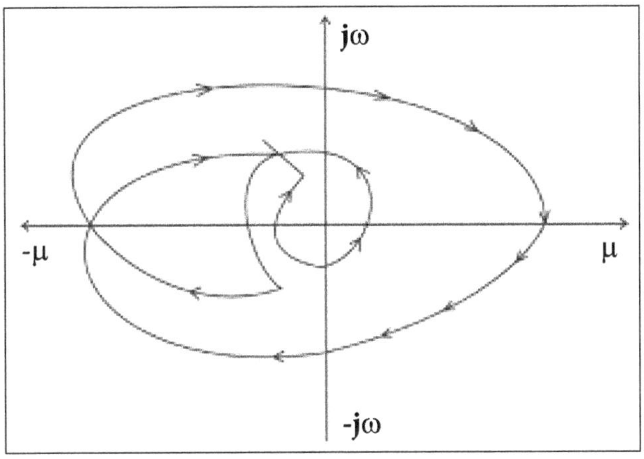

Step 6: Stability analysis: The Nyquist contour in G(s) H(s)–plane does not encircle the point (-1 +jo) because the open loop transfer function has no poles on the right half of the s – plane.

5

Classical Control Design Techniques

5.1 Lag, Lead and Lag-Lead Compensators

Basic Compensators using electrical network and the transfer functions:

Electrical Lead Network

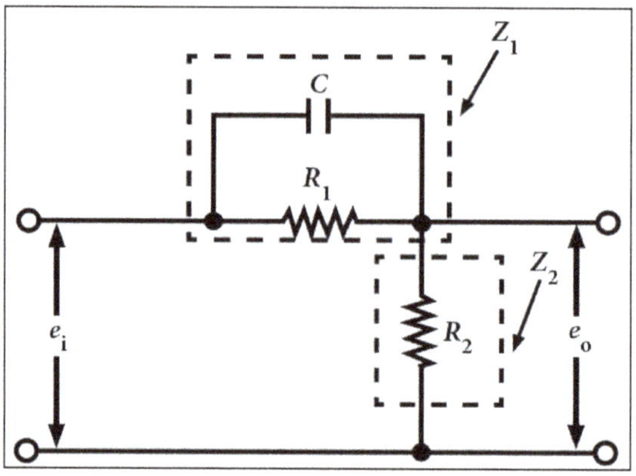

Lead Network.

Where, in s domain, the impedance are written as,

$$z_1(s)=R_1 \left\| \frac{1}{C_s} = \frac{R_1 * \frac{1}{C_s}}{R_1+\frac{1}{C_s}} = \frac{R_1}{R_1 C_s+1} \right.$$

$$z_2(s)=R_2$$

$$\frac{E_o(s)}{E_i(s)}=\frac{z_2}{z_1+z_2}=\frac{R_2}{\frac{R_1}{R_1 C_s+1}+R_2}$$

$$\frac{E_o(s)}{E_i(s)}=\frac{R_2(R_1 C_s+1)}{R_1+R_2(R_1 Cs+1)}=\frac{R_2(R_1 C_s+1)}{R_1+R_2+R_2 R_1 Cs}=\frac{R_2}{R_1+R_2}\times\frac{R_1 Cs+1}{1+\frac{R_2}{R_1+R_2}R_1 C_s}$$

$$\frac{E_o(s)}{E_i(s)} = \alpha \frac{1+sT}{1+s\alpha\alpha}, \text{for } 0<\alpha<1;$$

Where,

$$\alpha = \frac{R_2}{R_1+R_2} \text{ and } T = R_1 C$$

Electrical Lag Network

The impedances are written as,

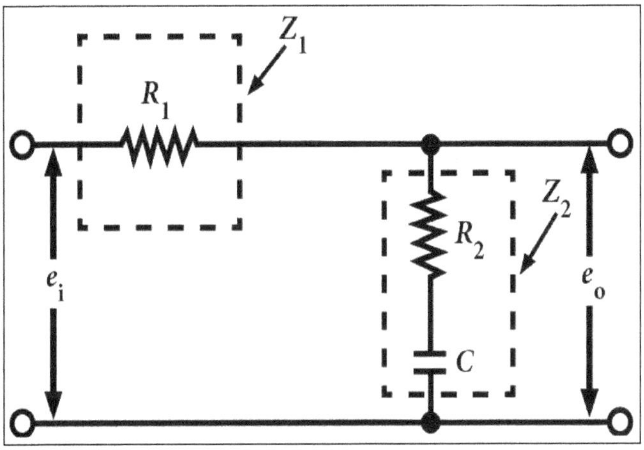

Lag Network.

$$Z_2(s) = R_2 + \frac{1}{C_s}$$

$$Z_1(s) = R_1$$

Transfer function,

$$\frac{E_o(s)}{E_i(s)} = \frac{Z_2}{Z_1+Z_2} = \frac{R_2 + \frac{1}{Cs}}{R_1 + R_2 + \frac{1}{Cs}}$$

$$\frac{E_o(s)}{E_i(s)} = \frac{R_2 Cs + 1}{(R_1+R_2)Cs + 1}$$

Multiplying and dividing R_2 in denominator gives,

$$\frac{E_o(s)}{E_i(s)}=\frac{R_2Cs+1}{\frac{(R_1+R_2)}{R_2}R_2Cs+1}$$

$$\frac{E_o(s)}{E_i(s)}=\frac{1+sT}{1+s\beta T}, \text{ for } \beta>1;$$

Where,

$$\beta=\frac{R_1+R_2}{R_2} \text{ and } T=R_2C$$

Electrical Lag-lead Network

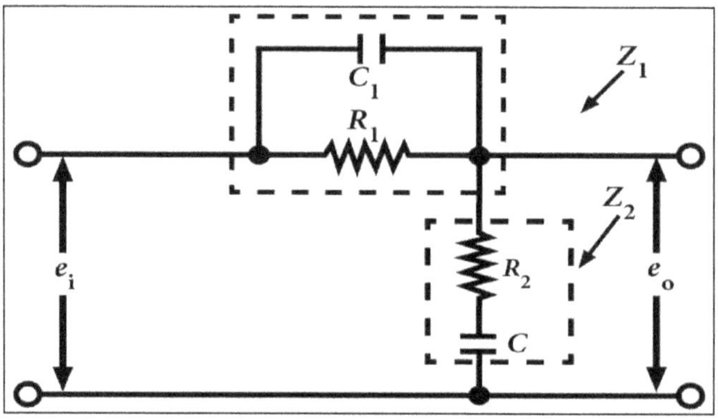

Lag-Lead Network.

Where in s-domain, the impedances are written as,

$$Z_2(s)=R_2+\frac{1}{Cs}$$

$$z_1(s)=R_1\left\|\frac{1}{Cs}=\frac{R_1*\frac{1}{Cs}}{R_1+\frac{1}{Cs}}\right.$$

Transfer function, $\frac{E_o(s)}{E_i(s)}=\frac{Z_2}{Z_1+Z_2}$

The transfer function of an electrical lag-lead network is given as,

$$\frac{E_o(s)}{E_i(s)}=\frac{1+sT_2}{1+s\beta T_2}\frac{1+sT_1}{1+s\alpha T_1}, \text{for } \alpha\beta=1;$$

Where,

$$\beta = \frac{R_1 + R_2}{R_2} \ \& \ \alpha = \frac{R_2}{R_1 + R_2} \text{ and } T_1 = R_1C \text{ and } T_2 = R_2C$$

Problems

1. Let us design a compensating network for $G(s) = \frac{K}{s(1+0.2s)(1+0.01s)}$ so that its phase margin at least will be 40° and steady state error will be in the final position will not exceed 2% of the final velocity of 50m/sec.

Solution:

Given:

$$G(s) = \frac{K}{s(1+0.2s)(1+0.01s)}$$

Phase margin at least = 40°

Steady state error in the final position will not exceed 2% of the final velocity of 50m/sec.

Formula to be used:

$$K_v = \lim_{s \to 0} sG(s)$$

Calculating the K from the velocity constant,

$$K_v = \lim_{s \to 0} sG(s) = \lim_{s \to 0} s \frac{k}{s(1+0.2s)(1+0.01s)} = k$$

$$k = 50$$

Drawing the bode plot for G (jw).

$$G(jw) = \frac{50}{jw(1+0.2jw)(1+0.01jw)}$$

Corner frequencies are,

$$w_{c1} = 5 \text{ red/sec}$$

$$w_{c2} = 100 \text{ red/sec,}$$

Choose w_l = 0.1 red/sec, w_h = 200 red/sec.

Magnitude Table

Term	Corner Frequency	Slope	Change in Slope
50/jw	--	-20dB/dec	
1/(1+j0.2w)	5 r ad/sec	-20 dB/dec	-20-20 = -40dB/dec
1/(1+j0.01w)	100 rad/sec	-20 dB/dec	-40-20 = -60dB/dec

Obtaining the Magnitude Calculation,

$$|M|_{wL} = 20\log\left|\frac{50}{jw}\right| = 20\log\left|\frac{50}{0.1}\right| = 54\,dB$$

$$|M|_{W_{cl}} = \text{Changeing slope from } W_{cl} \text{ to } W_L * \log\left(\frac{W_{cl}}{W_L}\right) + |M|_{W_L}$$

$$= -20\text{ dB} * \log\left(\frac{5}{0.1}\right) + 54\,dB$$

$$= 20\,dB$$

$$|M|_{W_{c2}} = \text{Changeing slope from } W_{c2} \text{ to } W_{cl} * \log\left(\frac{W_{c2}}{W_{cl}}\right) + |M|_{W_{cl}}$$

$$= -40\text{ dB} * \log\left(\frac{100}{5}\right) + 20\,dB$$

$$= -32\text{ dB}$$

$$|M|_{W_h} = \text{Changeing slope from } W_H \text{ to } W_{c2} * \log\left(\frac{W_H}{W_{c2}}\right) + |M|_{W_{c2}}$$

$$= -60\text{ dB} * \log\left(\frac{200}{100}\right) - 32\,dB$$

$$= -50\,dB$$

Obtaining Bode phase plot

$$\angle G(jw) = -90 - \tan^{-1}(0.2w) - \tan^{-1}(0.01w)$$

Phase Table

w	0.1	1	5	10	100	200
G(jw)	-91.2°	-101°	-138°	-159°	-222°	-242°

From the bode plot, W_{gc}=15 rad/sec, W_{pc} = 22 rad/sec & PM =10°

$$\phi_{max} = \text{Specified P.M} - \text{P.M of uncompen sated system at } W_{gc} + \zeta(5^\circ \text{ to } 12^\circ)$$

Obtain the attenuation,

$$\alpha = \frac{1-\sin\phi_m}{1+\sin\phi_m} = \frac{1-0.5735}{1+0.5735} = 0.271$$

The magnitude of the lead compensator at the frequency, at which the maximum phase occurs. i.e., W_{max} can be determined from magnitude corresponds to the frequency of

$$-20 * \log\frac{1}{\sqrt{\alpha}} = -5.67\,\text{dB}.$$

If we select this frequency 22rad/sec as the new P.M frequency, the uncompensated system's magnitude at this frequency must be -5.67dB, so that the magnitude at Wmax is 0dB.

From the bode plot, we see that W_{max} = 22rad/sec corresponds to -5.67dB.

The lead compensator corner frequencies are given by,

The zero of lead compensator,

The pole of lead compensator, $\frac{1}{T} = \sqrt{\alpha}w_{max} = \sqrt{0.271} * 22 = 11.4$

The T.F of compensator is given by, $\frac{1}{\alpha T} = \frac{w_{max}}{\sqrt{\alpha}} = \frac{22}{\sqrt{0.271}} = 42.2$

The T.F of compensator is given by,

$$G_c(s) = \frac{1}{\alpha}\frac{s+\frac{1}{T}}{s+\frac{1}{\alpha T}} = 3.96 \times \frac{s+11.4}{s+42.2} = 3.69\left[\frac{s+11.4}{s+42.2}\right]$$

5.2 Design of Compensators: Using Bode Plots

Phase Lead Compensator

From the frequency response of a simple PD controller, it is evident that the magnitude of the compensator continuously grows with increase in its frequency. The above feature is undesirable because it amplifies the high frequency noise which is typically present in any real system. In the lead compensator, a first order pole is added to the denominator of the PD controller at frequencies well higher than the corner frequency of the PD controller.

A typical lead compensator has the following transfer function.

$$C(s)=K\frac{\tau S+1}{\alpha \tau S+1},$$

Where, $\alpha < 1$ $\frac{1}{\alpha}$ is the ratio between the pole zero break point (corner) frequencies.

Magnitude of the lead compensator is $K\frac{\sqrt{1+\omega^2\tau^2}}{\sqrt{1+\alpha^2\omega^2\tau^2}}$ and the phase contributed by the lead compensator is given by,

$$\phi = \tan^{-1}\omega\tau - \tan^{-1}\propto\omega\tau$$

Therefore, a significant amount of phase is still provided with much less amplitude at high frequencies.

The frequency response of the typical lead compensator is shown in the below figure where the magnitude varies from $20\log_{10} K$ to $20\log_{10} K/\alpha$ and maximum phase is always less than 90° (around 60° in general).

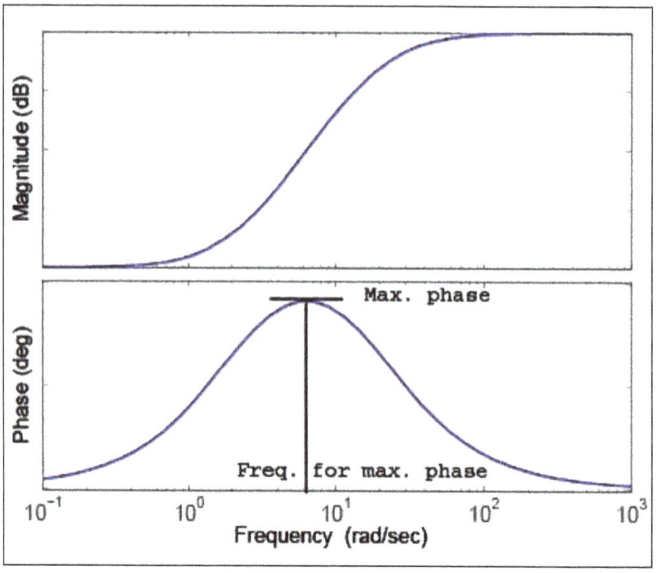

Frequency response of the lead compensator.

Frequency where the phase is maximum is given by,

$$\omega_{max} = \frac{1}{\tau\sqrt{\alpha}}$$

The maximum phase corresponds to,

$$\sin\phi_{max} = \frac{1-\alpha}{1+\alpha}$$

$$\Rightarrow \alpha = \left(\frac{1 - \sin(\phi_{max})}{1 + \sin(\phi_{max})} \right)$$

The magnitude of C(s) at ω_{max} is $\frac{K}{\sqrt{\alpha}}$.

Phase Lag Compensator

The essential feature of a lag compensator is to provide an increase d low frequency gain, thus decreasing the steady state error, without changing the transient response significantly. For frequency response design, it is easy to use the following transfer function of the lag compensator.

$$C_{lag}(s) = \alpha \frac{\tau s + 1}{\alpha \tau s + 1},$$

Where,

$$\alpha > 1$$

The above expression is only the lag part of the compensator. The overall compensator is $C(s) = KC_{lag}(s)$.

$$\text{When, } s \to 0,\ C_{lag}(s) \to \alpha$$
$$\text{When, } s \to \infty,\ C_{lag}(s) \to 1$$

Typical objective of lag compensator design is to provide the additional gain of α in the low frequency region and to leave the system with sufficient phase margin. The frequency response of the lag compensator with α = 4 and τ = 3 is shown in the below figure where the magnitude varies from $20\log_{10} \alpha$ d B to 0 dB.

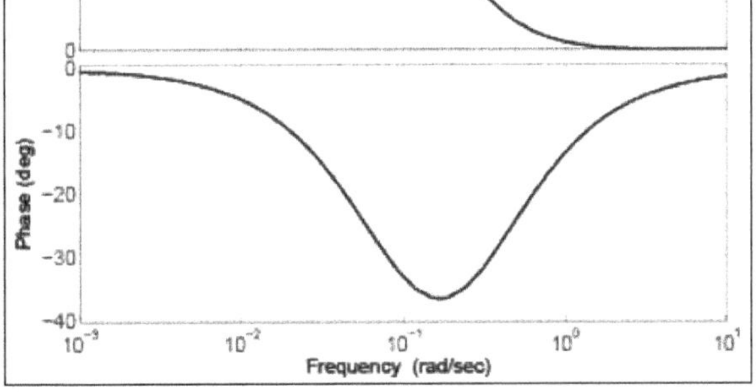

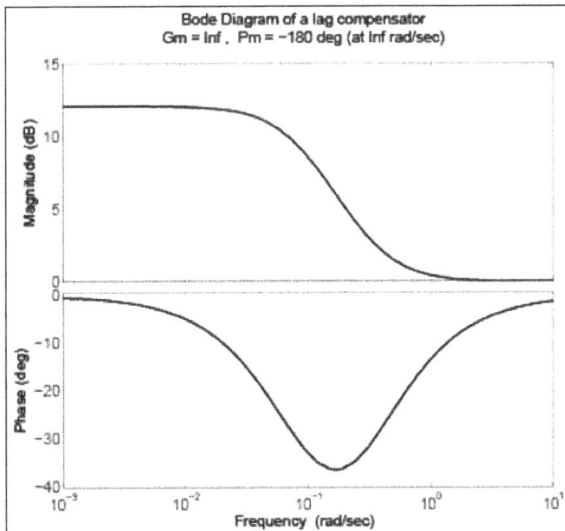

Frequency response of a lag compensator.

Since the lag compensator provides maximum lag near two corner frequencies, to maintain the PM of the system, zero of the compensator is chosen such that $\omega = 1/\tau$ should be much lower than the gain crossover frequency of the uncompensated system.

In general τ is designed such that $1/\tau$ is at least one decade below the gain cross - over frequency of the uncompensated system.

Problems

1. Consider the following system $G(s)=\frac{1}{s(s+1)}$, $H(s)=1$. Let us design a cascade lead compensator so that the phase margin (PM) is at least 45° and steady state error for a unit ramp input is ≤ 0.1.

Solution:

Given:

$$G(s)=\frac{1}{s(s+1)},\ H(s)=1$$

Phase margin (PM) = at least 45°

Steady state error for a unit ramp input ≤ 0.1

Formula to be used:

$$C(s)=K\frac{\tau s+1}{\alpha\tau s+1},$$

Where,

$$\alpha < 1$$

Steady state error for unit ramp input is,

$$\frac{1}{\lim_{s\to 0} sC(s)G(s)}$$

$$\omega_{max}=\omega_{g_{new}}=\frac{1}{\tau\sqrt{\alpha}}$$

$$G(j\omega)=\frac{1}{j\omega(j\omega+1)}$$

$$\text{Mag.}=\frac{1}{\omega\sqrt{1+\omega^2}}$$

$$\text{Phase} = -90° - \tan^{-1}\omega$$

$$\alpha = \left(\frac{1 - \sin(\phi_{max})}{1 + \sin(\phi_{max})}\right)$$

The lead compensator is,

$$C(s) = K\frac{\tau s + 1}{\alpha\tau s + 1},$$

Where,

$$\alpha < 1$$

When,

$$s \to 0,\ C(s) \to K.$$

Steady state error for unit ramp input is,

$$\frac{1}{\lim_{s\to 0} sC(s)G(s)} = \frac{1}{C(0)} = \frac{1}{K}$$

Thus,

$$\frac{1}{K} = 0.1, \text{ or } K = 10.$$

PM of the closed loop system is 45°. Let the gain cross - over frequency of the uncompensated system with K be ω_g.

$$G(j\omega) = \frac{1}{j\omega(j\omega + 1)}$$

$$\text{Mag.} = \frac{1}{\omega\sqrt{1 + \omega^2}}$$

$$\text{Phase} = -90° - \tan^{-1}\omega$$

$$\Rightarrow \frac{10}{\omega_g\sqrt{1 + \omega_g^2}} = 1$$

$$\frac{100}{\omega_g^2\left(1 + \omega_g^2\right)} = 1$$

$$\Rightarrow \omega_g = 3.1$$

Phase angle at ω_g = 3.1 i s -90 - $\tan^{-1}$ 3.1 = - 162°. Thus the PM of the uncompensated system with K is 18°.

If it was possible to add a phase without altering the magnitude, the additional phase lead required to maintain PM = 45° is 45° - 18° = 27° at ωg = 3.1 rad/sec. However, maintaining the same low frequency gain and adding a compensator would increase the crossover frequency. As a result of this, the actual phase margin will deviate from the designed one. Hence, it is safe to add the safety margin of ε to the required phase lead so that if it deviates also, still the phase requirement is met. General ly, ε is chosen between 5° to 15°. So the additional phase requirement is 27° + 10° = 37°. The lead part of the compensator will provide this additional phase at ω_{max}.

Thus,

$$\phi_{max} = 37^\circ$$

$$\Rightarrow \alpha = \left(\frac{1-\sin(\phi_{max})}{1+\sin(\phi_{max})}\right) = 0.25$$

The only parameter left to be designed is τ. To determine the value of τ, we should locate the frequency at which the uncompensated system has a logarithmic magnitude of

$$-20\log_{10}\frac{1}{\sqrt{\alpha}}.$$

Select this frequency as the new gain crossover frequency since the compensator provides again of $20\log_{10}\frac{1}{\sqrt{\alpha}}$ at ω_{max}. Thus,

$$\omega_{max} = \omega_{g_{new}} = \frac{1}{\tau\sqrt{\alpha}}$$

In this case, $\omega_{max} = \omega_{g_{new}} = 4.41$.

Thus,

$$\tau = \frac{1}{4.41\sqrt{\alpha}} = 0.4535$$

The lead compensator is thus given by,

$$C(s) = 10\frac{0.4535s+1}{0.1134s+1}$$

With this compensator, actual phase margin of the system becomes 49.6° which meets the design criteria. The corresponding Bode plot is shown in the figure.

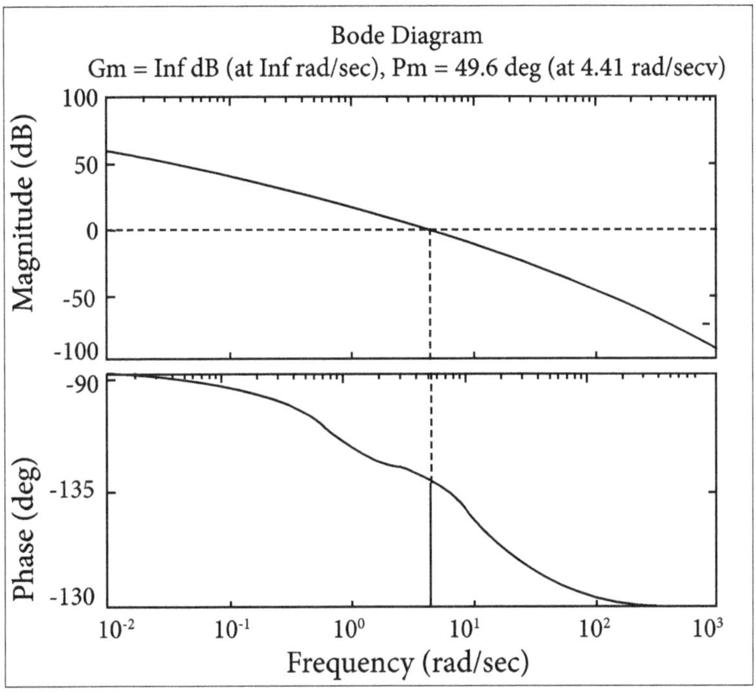

Bode plot of the compensated system.

2. Consider that the system $G(s)=\frac{1}{s(s+1)}, H(s)=1$ is subject to a sampled data control system with sampling time T = 0.2 sec. Let us draw the bode plot of the compensated and uncompensated system.

Solution:

Given:

$$G(s)=\frac{1}{s(s+1)}, H(s)=1$$

$$T = 0.2 \text{ sec}$$

Formula to be used:

$$C(\omega)=K\left(\frac{1+\tau\omega}{1+\alpha\tau\omega}\right) \qquad 0<\alpha<1$$

$$z=\frac{1+\omega T/2}{1-\omega T/2}$$

$$\omega_{g_{new}} = \omega_{max} = \frac{1}{\tau\sqrt{\alpha}}$$

$$K_v = \lim_{\omega \to 0} \omega C(\omega) G_\omega(\omega)$$

$$G_z(z) = (1 - z^{-1}) Z\left[\frac{1}{s^2(s+1)}\right]$$

$$= \frac{0.0187z + 0.0175}{z^2 - 1.8187z + 0.8187}$$

The bilinear transformation,

$$z = \frac{1 + \omega T/2}{1 - \omega T/2} = \frac{(1 + 0.1\omega)}{(1 - 0.1\omega)}$$

Will transfer $G_z(z)$ into ω -plane as,

$$G_\omega(\omega) = \frac{\left(1 + \frac{\omega}{300}\right)\left(1 - \frac{\omega}{10}\right)}{\omega(\omega + 1)}$$

First, we need to design a phase lead compensator so that PM of the compensated system is at least 50° with $Kv = 2$. The compensator in the ω -plane is given by,

$$C(\omega) = K\frac{1 + \tau\omega}{1 + \alpha\tau\omega} \qquad 0 < \alpha < 1$$

Design steps are as follows:

Step 1: K has to be found out from the Kv requirement.

Step 2: Make $\omega_{max} = \omega_{g_{new}}$.

Step 3: Compute the gain crossover frequency ωg and phase margin of the uncompensated system after introducing K in the system.

Step 4: At ω_g check the additional/required phase lead, add safety margin, and find out ϕ_{max}. Calculate α from the required ϕ_{max}. Since the lead part of the compensator provides a gain of $20 \log_{10} \frac{1}{\sqrt{\alpha}}$, find out the frequency where the logarithmic magnitude is $-20 \log_{10} \frac{1}{\sqrt{\alpha}}$. This is the new gain crossover frequency where the maximum phase lead should occur.

Step 5: Calculate τ from the relation $\omega_{g_{new}} = \omega_{max} = \frac{1}{\tau\sqrt{\alpha}}$.

Step 6: Now,

$$K_v = \lim_{\omega \to 0} \omega C(\omega) G_\omega(\omega) = 2$$

$$\Rightarrow K = 2$$

Step 7: Using MATLAB command "margin", phase margin of the system with K= 2 is computed as 31.6° with ω_g =1.26 rad/sec, as shown in the below figure:

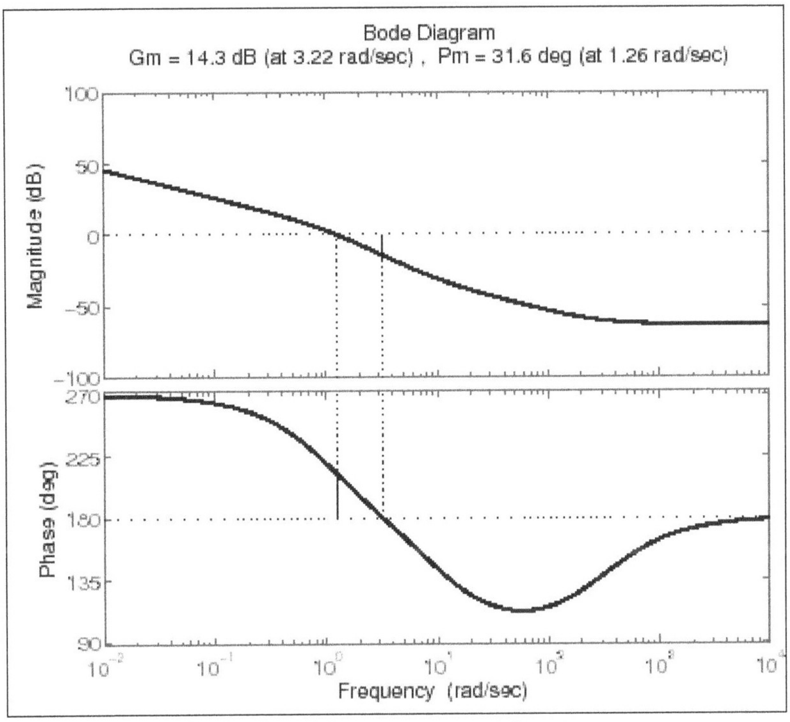

Bode plot of the uncompensated system.

Thus the required phase lead is 50° - 31.6° = 18.4°. After adding a safety margin of 11.6°, becomes 30°. Hence,

$$\alpha = \left(\frac{1 - \sin(29°)}{1 + \sin(29°)} \right) = 0.33$$

From the frequency response of the system, it can be found out that at ω =1.75 rad/sec, the magnitude of the system is $-20\log_{10}\frac{1}{\sqrt{\alpha}}$. Thus,

$$\omega_{max} = \omega_{gnew} = 1.75 \text{ rad/sec}.$$

This gives,

$$1.75 = \frac{1}{\tau\sqrt{\alpha}}$$

or,

$$\tau = \frac{1}{1.75\sqrt{0.33}} = 0.99$$

Thus the controller in ω-plane is $C(\omega) = 2\frac{1+0.99\omega}{1+0.327\omega}$.

The Bode plot of the compensated system is shown i n the below figure.

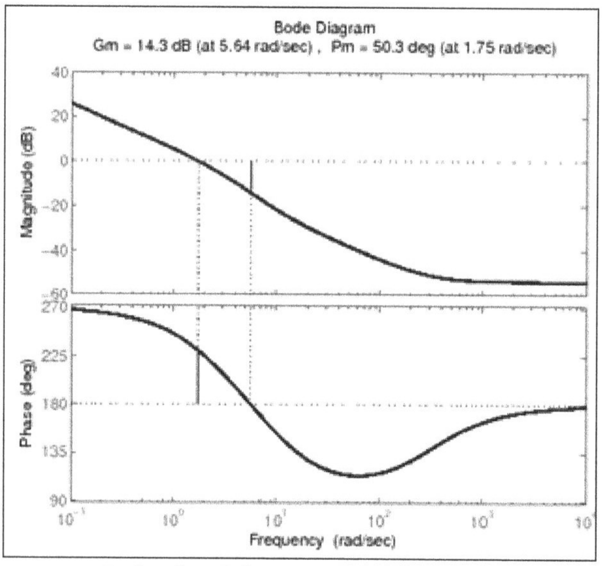

Bode plot of the compensated system.

Re-transforming the above controller into z-plane using the relation $\omega = 10\frac{z-1}{z+1}$, we get the controller in z -plane as $C_z(z) \cong 2\frac{2.55_z - 2.08}{z-0.53}$.

3. Consider the following system $G(s) = \frac{1}{(s+1)(0.5s+1)}$, $H(s) = 1$. Let us design a lag compensator so that the phase margin (PM) is at least 50° and steady state error to a unit step input is ≤0.1.

Solution:

Given:

$$G(s) = \frac{1}{(s+1)(0.5s+1)}, H(s) = 1$$

Phase margin (PM) = at least 50°

Steady state error to a unit step input ≤ 0.1

Formula to be used:

$$C(s)=KC_{lag}(s)=K\alpha\frac{\tau s+1}{\alpha\tau s+1},$$

Where,

$$\alpha > 1$$

Steady state error for unit ramp input is given by,

$$\frac{1}{\lim_{s\to o} sC(s)G(s)}$$

The overall compensator is given by,

$$C(s)=KC_{lag}(s)=K\alpha\frac{\tau s+1}{\alpha\tau s+1},$$

Where,

$$\alpha > 1$$

When,

$$s\to o,\ C(s)\to K\alpha.$$

Steady state error for unit step input is given by,

$$\frac{1}{1+\lim_{s\to o} C(s)G(s)}=\frac{1}{1+C(o)}=\frac{1}{1+K\alpha}$$

Thus,

$$\frac{1}{1+K\alpha}=0.1, \text{or}, K\alpha=9.$$

Now let us modify the system transfer function by introducing K with the original system. Thus the modified system becomes,

$$G_m(s)=\frac{K}{(s+1)(0.5s+1)}$$

PM of the closed loop system should be 50°. Let the gain crossover frequency of the uncompensated system with K be ω_g.

$$G_m(j\omega) = \frac{K}{(j\omega+1)(0.5j\omega+1)}$$

$$\text{Mag.} = \frac{K}{\sqrt{1+\omega^2}\sqrt{1+0.25\omega^2}}$$

$$\text{Phase} = -\tan^{-1}\omega - \tan^{-1}0.5\omega$$

Required PM is 50°. Since the PM is achieved only by selecting K, it might be deviated from this value when the other parameters are also designed. Thus we put a safety margin of 5° to the PM which makes the required PM to be 55°.

$$\Rightarrow 180° - \tan^{-1}\omega_g - \tan^{-1}0.5\omega_g = 55°$$

or,

$$\tan^{-1}\frac{\omega_g + 0.5\omega_g}{1-0.5\omega_g^2} = 125°$$

or,

$$\tan^{-1}\frac{1.5\omega_g}{1-0.5\omega_g^2} = \tan 125° = -1.43$$

or,

$$0.715\omega_g^2 - 1.5\omega_g - 1.43 = 0$$

$$\Rightarrow \omega_g = 2.8\,\text{rad/sec}$$

In order to make ω_g = 2.8 rad/sec, gain crossover frequency of the modified system, magnitude at ω_g should be 1.

Thus,

$$\frac{K}{\sqrt{1+\omega_g^2}\sqrt{1+0.25\omega_g^2}} = 1$$

Putting the value of ω_g in the last equation, we get K = 5.1.

Thus,

$$\alpha = \frac{9}{K} = 1.76$$

The only parameter left to be designed is τ.

Since the desired PM is already achieved with gain K, We should place ω = 1/ τ such that it does not much effect the PM of the modified system with K. If we place 1/ τ as one decade below the gain crossover frequency, then:

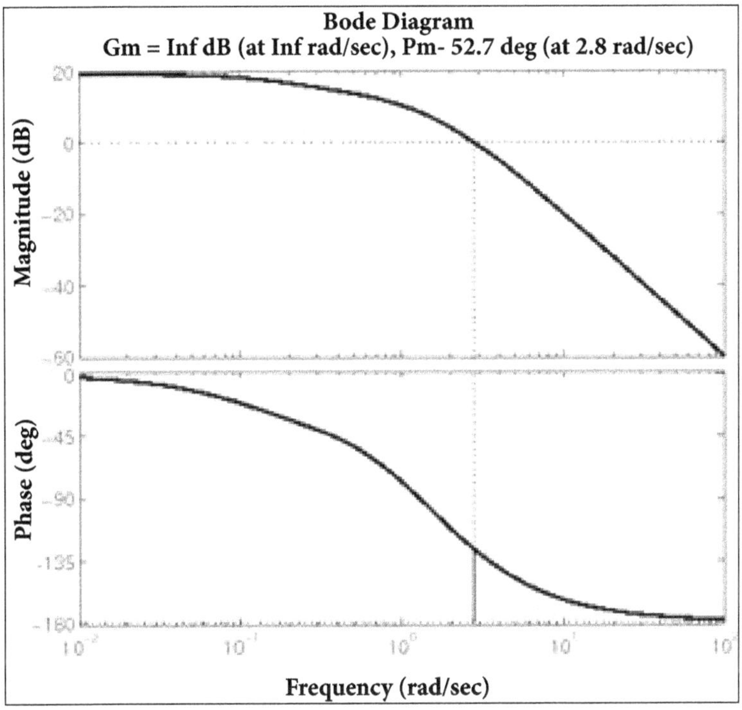

Bode plot of the compensated system.

$$\frac{1}{\tau} = \frac{28}{10},$$

or,

$$\tau = 3.57$$

The overall compensator is expressed as,

$$C(s) = 9\frac{3.57s + 1}{6.3s + 1}$$

With this compensator actual phase margin of the system becomes 52.7°, as shown in Figure b, which meets the design criteria.4. Now let us consider that the system $G(s) = \frac{1}{(s+1)(0.5s+1)}$, $H(s) = 1$ is subject to a sampled data control system with sampling time T = 0.1 sec. We would use MATLAB to derive the plant transfer function ω -plane. Let us draw the bode plot for the compensated and uncompensated system.

Solution:

Given:

$$G(s)=\frac{1}{(s+1)(0.5s+1)},H(s)=1$$

T = 0.1 sec

Use the below commands,

$$>> s = tf('s');$$

$$>> gc = 1/((s+1)*(0.5*s+1));$$

$$>> gz = c2d(gc,0.1,'zoh');$$

We get,

$$G_z(z)=\frac{0.009z+0.0008}{z^2-1.724z+0.741}$$

The bilinear transformation,

$$z=\frac{1+\omega T/2}{1-\omega T/2}=\frac{(1+0.05\omega)}{(1-0.05\omega)}$$

Will transfer G_z (z) into ω-plane. Let us u se the below commands,

$$>> aug = [0.1,1];$$

$$>> gwss = bilin(ss(gz),-1,'S\_Tust',aug)$$

$$>> gw = tf(gwss)$$

To find out the transfer function in ω-plane as,

$$G_\omega(\omega) = \frac{1.992-0.09461\omega-0.00023\omega^2}{\omega^2+2.993\omega+1.992}$$

$$\cong \frac{-0.00025(\omega-20)(\omega+400)}{(\omega+1)(\omega+2)}$$

The Bode plot of the uncompensated system is shown in the below figure:

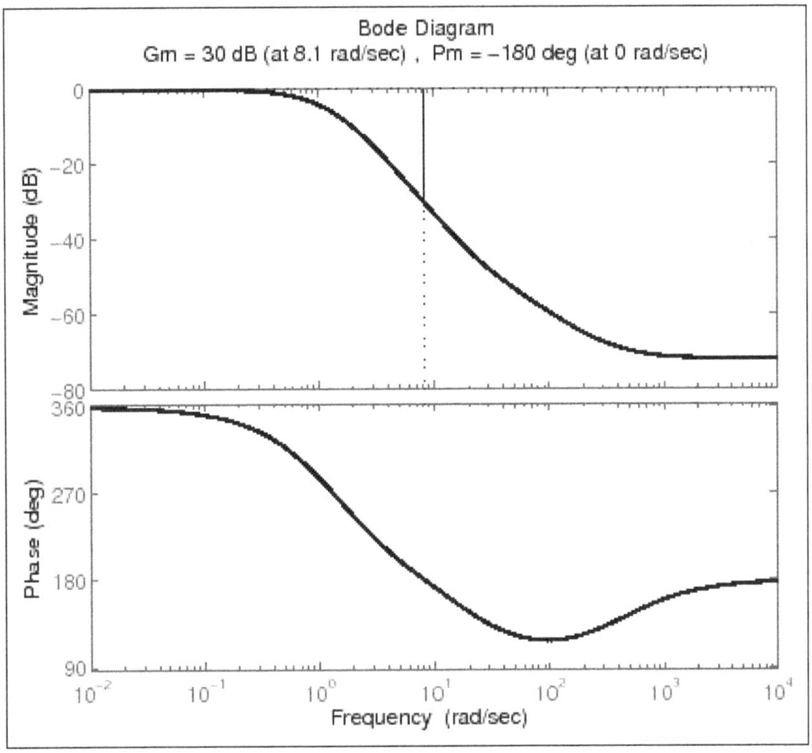

Bode plot of the uncompensated system.

We need to design a phase lag compensator so that PM of the compensated system is at least 50° and steady state error to a unit step input is 0.1. The compensator in ω -plane is,

$$C(\omega)=K\alpha\frac{1+\tau\omega}{1+\alpha\tau\omega} \qquad \alpha>1.$$

Where,

$$C(0)=K\alpha$$

Since,

$G_\omega(0) = 1$, $K\alpha = 9$, for 0.1 steady state error.

Now let us modify the system transfer function by introducing K to the original system.

Thus the modified system becomes $G_m(\omega)=\frac{-0.00025K(\omega-20)(\omega+400)}{(\omega+1)(\omega+2)}$.

PM of the closed loop system should be 50°. Let the gain crossover frequency of the uncompensated system with K be ωg.

Then,

$$\text{Mag.}(G_m)=\frac{0.00025K\sqrt{400+\omega^2}\sqrt{160000+\omega^2}}{\sqrt{1+\omega^2}\sqrt{4+\omega^2}}$$

$$\text{Phase}(G_m)=-\tan^{-1}\omega-\tan^{-1}0.5\omega-\tan^{-1}0.05\omega+\tan^{-1}0.0025\omega$$

Required PM is 50°. Let us put a safety margin of 5°. Thus the PM of the system modified with K should be 55°.

$$\Rightarrow 180°-\tan^{-1}\omega_g-\tan^{-1}0.5\omega_g-\tan^{-1}0.05\omega_g+\tan^{-1}0.0025\omega_g=55°$$

$$\tan^{-1}\frac{\omega_g+0.5\omega_g}{1-0.5\omega_g^2}-\tan^{-1}\frac{0.05\omega_g-0.0025\omega_g}{1+0.000125\omega_g^2}=125°$$

By solving the above, ω_g = 2.44 rad/sec. Thus the magnitude at ω_g should be 1.

$$\Rightarrow\frac{0.00025K\sqrt{400+\omega_g^2}\sqrt{160000+\omega_g^2}}{\sqrt{1+\omega_g^2}\sqrt{4+\omega_g^2}}=1$$

Putting the value of ωg in the last equation, we get K = 4.13.

Thus,

$$\alpha=\frac{9}{K}=2.18$$

If we place 1/ τ one decade below the gain crossover frequency, then

$$\frac{1}{\tau}=\frac{2.44}{10},\ \text{or}\ \tau=4.1$$

Thus the controller in ω -plane is,

$$C(\omega)=9\frac{1+4.1\omega}{1+8.9\omega}$$

Re-transforming the above controller into z -plane using the relation $\omega=20\frac{z-1}{z+1}$, we get

$$C_z(z)=9\frac{1+20\times 4.1\times\frac{z-1}{z+1}}{1+20\times 8.9\times\frac{z-1}{z+1}}$$

$$=9\frac{83z-81}{179z-177}$$

The Bode plot of the compensated system is shown in the below figure:

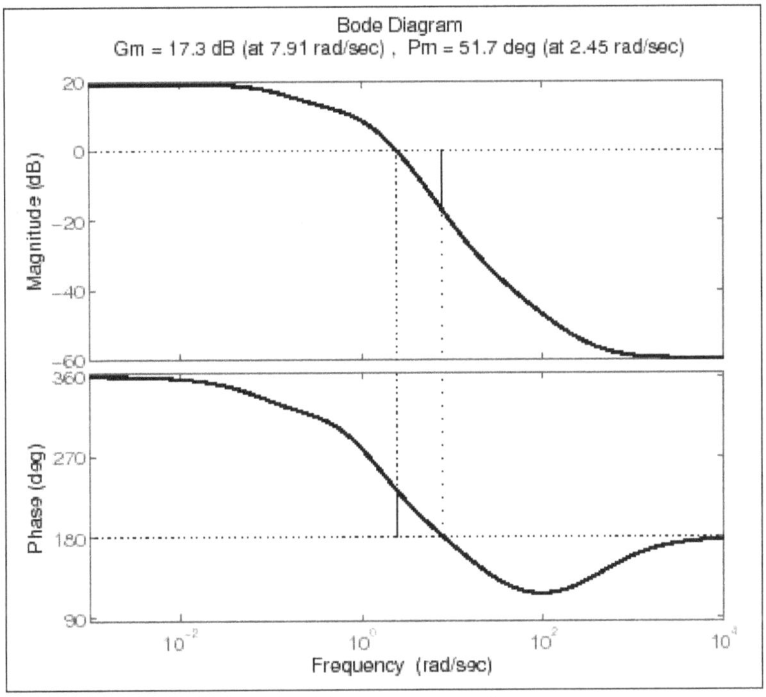

Bode plot of the compensated system.

6

State Space Analysis of LTI Systems

6.1 Concepts of State, State Variables and State Model

State

The state of dynamic system is defined as a minimal set of variables such that the knowledge of these variables at t = to together with the knowledge of the inputs for $t \geq t_o$, completely determines the behavior of the system for $t > t_o$.

State Variable

The variables involved in determining the state of a dynamic system x (t), are called as the state variables. $X_1(t)$, $X_2(t)$,..... X_n (t) are called state variables. They are normally energy storing elements contained in the system.

Modal Matrix

The modal matrix M can be formed from Eigen vectors. The modal matrix M is obtained by $M = [m_1, m_2, m_3 \ldots m_n]$.

Need of state variables:

- The state variables can be utilized for the purpose of feedback.
- Implementation of design becomes straight forward.
- The solution of state equation gives time variation of variables, which have direct relevance to the physical system.

State Space Description for Continuous LTI Systems

Input-output form can be the differential equation of higher order. A "simpler" description is a system of first order differential equations.

X_i (t), i = 1,...., n internal variables (states).

$$x = \begin{bmatrix} x_1(t) \\ \vdots \\ x_n(t) \end{bmatrix} \text{ state vector;}$$

$$u=\begin{bmatrix} u_1(t) \\ \vdots \\ u_n(t) \end{bmatrix} \text{input vector;}$$

$$\dot{x}(t)=A(t)x(t)+B(t)u(t)$$

$$y(t)=C(t)x(t)+D(t)u(t)$$

$A \in R^{nxn}$, $B \in R^{nxm}$, $C \in R^{pxn}$, $D \in R^{pxm}$

If A, B, C, D are constant matrices, then the model is linear and time-invariant.

$$\dot{x}(t)=Ax(t)+Bu(t)$$

$$y(t)=Cx(t)+Du(t)$$

- Set initial conditions $x_o=x(t_o)$
- Set input signal $u(t)$, $t=(t_o,\infty)$

The solution can be calculated analytically as,

$$x(t)=e^{A(t-t_o)}x(t_o)+\int_{t_o}^{t} e^{A(t-\tau)}Bu(\tau)d\tau$$

$$y(t)=Ce^{A(t-t_o)}x(t_o)+\int_{t_o}^{t} Ce^{A(t-\tau)}Bu(\tau)d\tau\ Du(t) \text{Convolution equestion}$$

Matrix exponential $\Rightarrow e^{At}$

6.1.1 State Space Representation of Transfer Function

State Equation

State equation relates the state variable and the inputs and is given by,

X (t) = AX + BU

The vector X is called $n \times 1$ state vector and U is $m \times 1$ input vector, A is an n x m square matrix, B is an n × m matrix.

Transfer Function from State Variable Representation

The transfer function of a single input-single output (SISO) system can be obtained from the state variable equations,

$$\dot{x} = Ax + Bu$$

$$y = Cx$$

Where y is the single output and u is the single input.

The Laplace transform of the equations can be expressed as,

$$sX(s) = AX(s) + BU(s)$$

$$Y(s) = CX(s)$$

Where, B is an nx1 matrix, since u is the single input. We do not include the initial conditions, since we seek the transfer function. Reordering the equation, we get,

$$[sI - A]X(s) = BU(s)$$

$$X(s) = [sI - A]^{-1} BU(s) = \phi(s)BU(s)$$

$$Y(s) = C\phi(s)BU(s)$$

Therefore, the transfer function G(s) = Y(s)/U(s) is given by,

$$G(s) = C\phi(s)B$$

Solutions of the State Equations

Consider the state equation of the linear time invariant system as,

$$X(t)AX(t) + BU(t)$$

The matrices A and B are constants.

This state equation can be of two types:

- Homogeneous.
- Non-homogeneous.

Homogeneous Equation

If A is a constant matrix and the input control forces are zero, then the equation takes the form,

$$\dot{x}(t) = A\,X(t)$$

Such an equation is known as homogeneous equation.

In such systems the driving force is provided by the initial conditions of the system to produce its output.

For example, consider a series RC circuit in which the capacitor is initially charged to V volts. The output is the current. Now, there is no input control force i.e. the external voltage that is applied to the system. But, the initial voltage that is applied on the capacitor drives the current through the system and the capacitor starts to discharge through the resistance R. Such a system which works on the initial conditions without any input applied to it is called as homogeneous system.

Non-homogeneous Equation

If A is a constant matrix and matrix U (t) is a non-zero vector i.e. the input control forces are applied to the system, then the equation takes normal form as,

$$\dot{x}(t)=AX(t)+BU(t)$$

Such an equation is called non-homogeneous equation. Most of the practical systems require inputs to drive them. Such systems are termed as non-homogeneous linear systems.

The solution of the state equation is obtained by considering the basic method of finding the solution of homogeneous equation.

Advantages of state space representation:

- It can be applied to both linear and non-linear time variant / time invariant systems.
- N^{th} order differential equations can be expressed as 'n' equation of first order equation.
- It is a time domain approach.
- It is suitable for digital computer computation.

Problems

1. Let us determine the transfer function G(s) = Y(s)/U(s) for the RLC circuit as described by the state differential function,

$$\dot{x}=\begin{bmatrix} 0 & -\frac{1}{C} \\ \frac{1}{L} & -\frac{R}{L} \end{bmatrix}x+\begin{bmatrix} \frac{1}{C} \\ 0 \end{bmatrix}u,\ y=\begin{bmatrix} 0 & R \end{bmatrix}x$$

Solution:

Given:

$$G(s) = Y(s)/U(s)$$

$$\dot{x} = \begin{bmatrix} 0 & -\frac{1}{C} \\ \frac{1}{L} & -\frac{R}{L} \end{bmatrix} x + \begin{bmatrix} \frac{1}{C} \\ 0 \end{bmatrix} u, \quad y = \begin{bmatrix} 0 & R \end{bmatrix} x$$

Formula to be used:

$$\phi(s) = [sI - A]^{-1}$$

$$[sI - A] = \begin{bmatrix} s & \frac{1}{C} \\ -\frac{1}{L} & s + \frac{R}{L} \end{bmatrix}$$

$$\phi(s) = [sI - A]^{-1} = \frac{1}{\Delta(s)} \begin{bmatrix} s + \frac{R}{L} & -\frac{1}{C} \\ \frac{1}{L} & s \end{bmatrix}$$

$$\Delta(s) = s^2 + \frac{R}{L}s + \frac{1}{LC}$$

Then the transfer function is expressed as,

$$G(s) = \begin{bmatrix} 0 & R \end{bmatrix} \begin{bmatrix} \frac{s + \frac{R}{L}}{\Delta(s)} & -\frac{1}{C\Delta(s)} \\ \frac{1}{L\Delta(s)} & \frac{s}{\Delta(s)} \end{bmatrix} \begin{bmatrix} \frac{1}{C} \\ 0 \end{bmatrix}$$

$$G(s) = \frac{R/LC}{\Delta(s)} = \frac{R/LC}{s^2 + \frac{R}{L}s + \frac{1}{LC}}$$

6.2 Diagonalization Techniques

The physical-variable state model is not convenient for investigation of system properties and evaluation of time-response.

The diagonal canonical form or normal form of state model, in the matrix a turns out to be a diagonal matrix, is most suitable for this purpose. Thus, it is useful to study the techniques by means of which a general state model can be transformed into a diagonal canonical form. These techniques are often referred to as diagonalization techniques.

Consider a multi-input-multi-output state model,

$$x(t) = Ax(t) + Bu(t)$$

$$y(t) = Cx(t) + Du(t)$$

In the state model given by equations above the matrix A is non-diagonal. Let us define a new state-vector z (t) such that;

$$x(t) = Mz(t)$$

Where,

M is an n * n non-singular, constant matrix.

Under this similarity transformation, the original state model equations modifies to,

$$\dot{z}(t) = M^{-1}AMz(t) + M^{-1}Bu(t)$$

$$y(t) = CMz(t) + Du(t)$$

If we select the matrix M such that $M^{-1}AM$ is a diagonal zed matrix of matrix A then the model given by equations just above is a state model in diagonal canonical form. Under this condition, the matrix M is called the diagonal zing matrix or the modal matrix and is constructed by placing the Eigen-vectors of matrix A together.

I.e., If $m_1, m_2, m_3, \ldots, m_n$ be the Eigen-vectors of matrix A corresponding to the Eigen-values $\lambda_1, \lambda_2, \lambda_3, \ldots, \lambda_n$ respectively, then the modal matrix is given by,

$$M = [m_1 \vdots m_2 \vdots m_3 \vdots \ldots \vdots m_n]n * n$$

Thus, the general state model equations modifies to a new state model (in diagonal canonical form), given by,

$$\dot{z}(t) = \Lambda z(t) + \tilde{B}u(t)$$

$$y(t)=\tilde{C}z(t)+Du(t)$$

The matrix A can be directly obtained without the need to compute M-1 AM, since the diagonal elements of matrix A are given by distinct Eigen-values λ_1, λ_2 , ... λ_n of matrix A.

$$\Lambda=\begin{bmatrix} \lambda_1 & 0 & 0 & \cdots & 0 \\ 0 & \lambda_2 & 0 & \cdots & 0 \\ 0 & 0 & \lambda_3 & \cdots & 0 \\ \vdots & \vdots & \vdots & & \vdots \\ 0 & 0 & 0 & \cdots & \lambda_H \end{bmatrix}_{H\times H} = M^{-1}AM$$

Note that A and A matrices have the same characteristic equation and therefore, the Eigen-values are invariant under the transformation.

If matrix A is given in Bush's or Phase-Variable or Companion Form, i.e.,

$$A=\begin{bmatrix} 0 & 1 & 0 & \cdots & 0 \\ 0 & 0 & 1 & \cdots & 0 \\ \vdots & \vdots & \vdots & & \vdots \\ 0 & 0 & 0 & \cdots & 1 \\ -a_n & -a_{n-1} & -a_{n-2} & \cdots & -a_1 \end{bmatrix}_{H\times H}$$

Then the modal matrix with reference to Bus h's or Phase-Variable or Companion Form of matrix A can be shown to be a special matrix, called the Vander Monde Matrix and is given by,

$$V=\begin{bmatrix} 1 & 1 & \cdots & 1 \\ \lambda_1 & \lambda_2 & \cdots & \lambda_n \\ \lambda_1^2 & \lambda_2^2 & \cdots & \lambda_n^2 \\ \vdots & \vdots & \cdots & \vdots \\ \lambda_1^{n-1} & \lambda_2^{n-1} & \cdots & \lambda_n^{n-1} \end{bmatrix}_{n\times n}$$

Where λ_1, λ_2 , ... λ_n are distinct Eigen-values of matrix A.

An advantage of diagonal matrix is that the inverse of such a matrix can be obtained merely by inspection, e.g.,

$$\begin{bmatrix} \alpha & 0 & 0 \\ 0 & \beta & 0 \\ 0 & 0 & \gamma \end{bmatrix}^{-1} = \begin{bmatrix} 1/\alpha & 0 & 0 \\ 0 & 1/\beta & 0 \\ 0 & 0 & 1/\gamma \end{bmatrix}$$

Now consider the case in which the matrix A involves multiple Eigen-values, it is impossible to be diagonal zed.

However, there exists a similarity transformation x (t) = Sz (t) such that the matrix J = S^{-1} AS is almost a diagonal matrix. The matrix J is called to be in Jordan Canonical Form. Under this similarity transformation, the state model in Jordan Canonical Form can be expressed as,

$$\dot{z}(t)=Jz(t)+\tilde{B}u(t)$$

$$y(t)=\tilde{C}z(t)+Du(t)$$

Where,

$$J = S^{-1}AS; \text{ A Jordan Matrix (Almost Diagonal)}$$

$$\tilde{B} = S^{-1}B$$

and,

$$\tilde{C} = CS$$

Problems

1. Consider a matrix A given by $A = -\begin{bmatrix} 0 & 1 & 0 \\ 3 & 0 & 2 \\ -12 & -7 & -6 \end{bmatrix}$. Let us compute diagonal matrix for A.

Solution:

Given:

$$A = -\begin{bmatrix} 0 & 1 & 0 \\ 3 & 0 & 2 \\ -12 & -7 & -6 \end{bmatrix}$$

Formula to be used:

$$|\lambda I - A| = 0$$

$$M = [m_1 \vdots m_2 \vdots m_3]$$

$$\Lambda = M^{-1}AM$$

Characteristic equation of matrix A is given by,

$$|\lambda I - A| = 0$$

Now,

$$\lambda I - A = \begin{bmatrix} \lambda & -1 & 0 \\ -3 & \lambda & -2 \\ 12 & 7 & \lambda+6 \end{bmatrix}$$

$$\therefore \ |\lambda I - A| = \begin{bmatrix} \lambda & -1 & 0 \\ -3 & \lambda & -2 \\ 12 & 7 & \lambda+6 \end{bmatrix} = 0$$

$$(\lambda+1)\ (\lambda+2)(\lambda+3) = 0$$

Therefore, the Eigen values of matrix A are obtained as,

$$\Lambda_1 = -1$$

$$\Lambda_2 = -2$$

$$\Lambda_3 = -3$$

Now the Eigen-vector m_1 associated with the Eigen-value $\lambda_1 = -1$ is obtained by solving the equation,

$$(\lambda_1 I - A) m_1 = 0$$

$$\begin{bmatrix} -1 & -1 & 0 \\ -3 & -1 & -2 \\ 12 & 7 & 5 \end{bmatrix} \begin{bmatrix} m_{11} \\ m_{21} \\ m_{31} \end{bmatrix} = 0$$

$$-m_{11} - m_{21} = 0$$

$$-3m_{11} - m_{21} - 2m_{31} = 0$$

$$12m_{11} + 7m_{21} + 5m_{31} = 0$$

Select $m_{11} = 1$

We get,

$$m_{21} = -1$$

or,

$$-2m_{12} - m_{22} = 0$$

$$-3m_{12} - 2m_{22} - 2m_{32} = 0$$

$$12m_{12} + 7m_{22} + 4m_{32} = 0$$

Select, $m_{12} = 2$

We get,

$$m_{22} = -4$$

and,

$$m_{32} = 1$$

Thus,

$$m_2 = \begin{bmatrix} m_{12} \\ m_{22} \\ m_{32} \end{bmatrix} = \begin{bmatrix} 2 \\ -4 \\ 1 \end{bmatrix}$$

Similarly, we can compute the Eigen vector m_3 associated with $\lambda_3 = -3$ as,

$$m_3 = \begin{bmatrix} m_{12} \\ m_{22} \\ m_{32} \end{bmatrix} = \begin{bmatrix} 1 \\ -3 \\ 3 \end{bmatrix}$$

The modal matrix M is obtained by,

$$M = [m_1 \vdots m_2 \vdots m_3]$$

$$M = \begin{bmatrix} 1 & 2 & 1 \\ -1 & -4 & -3 \\ -1 & 1 & 3 \end{bmatrix}$$

$$\text{Adj } M = \begin{bmatrix} +\begin{vmatrix} -4 & -3 \\ 1 & 3 \end{vmatrix} & -\begin{vmatrix} -1 & -3 \\ -1 & 3 \end{vmatrix} & +\begin{vmatrix} -1 & -4 \\ -1 & 1 \end{vmatrix} \\ -\begin{vmatrix} 2 & 1 \\ 1 & 3 \end{vmatrix} & +\begin{vmatrix} 1 & 1 \\ -1 & 3 \end{vmatrix} & -\begin{vmatrix} 1 & 2 \\ -1 & 1 \end{vmatrix} \\ +\begin{vmatrix} 2 & 1 \\ -4 & -3 \end{vmatrix} & -\begin{vmatrix} 1 & 1 \\ -1 & -3 \end{vmatrix} & +\begin{vmatrix} 1 & 2 \\ -1 & -4 \end{vmatrix} \end{bmatrix}^T$$

$$= \begin{bmatrix} -9 & 6 & -5 \\ -5 & 4 & -3 \\ -2 & 2 & -2 \end{bmatrix}^T = \begin{bmatrix} -9 & -5 & -2 \\ 6 & 4 & 2 \\ -5 & -3 & -2 \end{bmatrix}$$

$$|M| = \begin{vmatrix} 1 & 2 & 1 \\ -1 & -4 & -3 \\ -1 & 1 & 3 \end{vmatrix} = -2$$

$$M^{-1} = \frac{\text{Adj } M}{|M|} = -\frac{1}{2}\begin{bmatrix} -9 & -5 & -2 \\ 6 & 4 & 2 \\ -5 & -3 & -2 \end{bmatrix}$$

$$M^{-1} = \begin{bmatrix} 4.5 & 2.5 & 1 \\ -3 & -2 & -1 \\ 2.5 & 1.5 & 1 \end{bmatrix}$$

$$\therefore \qquad \Lambda = M^{-1} AM$$

$$= \begin{bmatrix} 4.5 & 2.5 & 1 \\ -3 & -2 & -1 \\ 2.5 & 1.5 & 1 \end{bmatrix} \begin{bmatrix} 0 & 1 & 0 \\ 3 & 0 & 2 \\ -12 & -7 & -6 \end{bmatrix} \begin{bmatrix} 1 & 2 & 1 \\ -1 & -4 & -3 \\ -1 & 1 & 3 \end{bmatrix}$$

$$= \begin{bmatrix} 4.5 & 2.5 & 1 \\ -3 & -2 & -1 \\ 2.5 & 1.5 & 1 \end{bmatrix} \begin{bmatrix} -1 & -4 & -3 \\ 1 & 8 & 9 \\ 1 & -2 & -9 \end{bmatrix} \begin{bmatrix} -1 & 0 & 0 \\ 0 & -2 & 0 \\ 0 & 0 & -3 \end{bmatrix}$$

$$\therefore \qquad e^{\Lambda t} = \begin{bmatrix} e^{-t} & 0 & 0 \\ 0 & e^{-2t} & 0 \\ 0 & 0 & e^{-3t} \end{bmatrix}$$

6.3 State Transition Matrix and its Properties

State Model of nth Order System

$$x(t) = A(t)x(t) + B(t)u(t)$$

$$y(t) = C(t)x(t) + D(t)u(t)$$

Where,

A (t) - state matrix; B (t) - input matrix

C (t) - output matrix; D (t) – direct trmn matrix

ϕ (t, τ) is called the state transition matrix.

Properties

$$\phi\,(t,t) = I,$$

$$\phi^{-1}(t,\tau) = \phi(\tau,t),$$

$$\phi(t_1\,,t_2) = \phi(t_1\,,t_o\,)\,\phi(t_o\,,t_2)$$

$$\frac{d}{dt}\phi(t,\tau) = A\,\phi(t,\tau),\ \phi\,(\tau,\tau) = I$$

Proof

$$\phi(t,t) = P(t)P^{-1}(t) = I$$

$$\phi^{-1}(t,\tau) = \left[P(t)P^{-1}(\tau)\right]^{-1} = I$$

$$\phi(t_1,t_2) = P(t_1)P^{-1}(t_2)P(t_2)P^{-1}(t_3) = P(t_1)P^{-1}(t_3) = \phi(t_1,t_3\,)$$

$$\frac{d}{dt}\phi(t,\tau) = \frac{d}{dt}P(t)P(\tau)^{-1} = A(t)P(t)P(\tau) = A(t)\phi(t,\tau)$$

6.3.1 Concepts of Controllability and Observability

Controllability and observability are two important properties of state models which are to be studied prior to designing a controller. Controllability deals with the possibility of forcing the system to a particular state by the application of control input. If a state is uncontrollable, then no input will be able to control that state.

Also the initial states can be observed from the output is determined using observability property. Hence, if the state is not observable, then the controller will not be able to determine its behavior from the system output and hence not be able to use that state to stabilize the system.

Concept of Controllability

A general n^{th} order multi-input linear time invariant system (X) = AX + BU, Y = CX is completely controllable if and only if the rank of the composite matrix,

$$Q_c = [B : AB \ldots\ldots A^{n-1}\ B] \text{ is n.}$$

Let us consider the Continuous Time system,

$$X = Ax + Bu,$$

$$y = Cx + Du.$$

Where,

A = n x n matrix, B = n x 1 matrix & C = 1 x n matrix.

State Controllability

It is completely state controllable if and only if the vectors $B,\ AB, A^2\ B,\ \ldots\ldots\ A^{n-1} B$ are linearly independent (or) $\left| B\ AB\ A^2\ B\ \ldots\ldots\ A^{n-1} B \right|$ is of rank n.

Output Controllability

Condition f or controllability is given by,

$$|C\ B\ C\ AB\ C\ A^2\ B \ldots\ldots C\ A^{n-1} B\ D| \text{ is of rank n.}$$

Observability:

$$X = Ax,$$

$$y = Cx.$$

Where,

A = n x n matrix, C = m x n matrix.

It is completely state observable if and only if n x mn matrix,

$$\left[C^*\ A^* C^* \bullet\bullet \left(A^*\right)^{n-1} C\right] (\text{or}) \begin{vmatrix} C \\ CA \\ CA^2 \\ \bullet \\ \bullet \\ CA^{n-1} \end{vmatrix} \text{ is of rank 'n' or has 'n'}$$

Linearly independent vectors.

Tests for Controllability

Theorem

For the continuous time system, the three tests for controllability of system is given by the following theorem:

- The system is controllable over the interval [0, T], for some T > 0.
- The system is controllable over any interval $[t_0, t_1]$ with t1 > t0.
- The reachability grammian,

$$W_{r,T} := \int_0^T \Phi(T,t)\ B(t)\ B(t)^T\ \Phi(T,t)^T\,dt$$

$$= \int_0^T e^{At}\,BB^*\,e^{A^*t}\,dt \text{ is full rank n.}$$

(This test works for time varying systems also).

- The controllability matrix is given by,

  $C = (B\ AB\ A^2\ B \ldots A^{n-1}\ B)$ has rank n.

Notice that the controllability matrix has dimension $n \times nm$ where m is the dimension of the control vector u.

- Belovich-Popov-Haut us test,

For each $s \in C$, the matrix (sI - A B) has rank n. Note that rank of (sI – A, B) is less than n, only if s is an eigenvalue of A.

The Cayley Hamilton theorem is considered to prove this result.

Cayley Hamilton Theorem

Let $A \in R\ n \times n$. Consider the polynomial $\Delta(s) = \det(sI - A)$,

Then,

$$\Delta(A) = 0.$$

Notice that if λ_i is an eigenvalue of A, then $\Delta(\lambda_i) = 0$.

Proof

This is true for any A, but is easy to see using the Eigen decomposition of A when A is semi-simple,

$$A = T\,\Lambda\,T^{-1}$$

Where Λ is diagonal with eigenvalues on its diagonal. Since $\psi\ (\lambda_i)=0$ by definition,

$$\Delta(A)=T\ \psi(\Lambda)T^{-1}=0$$

Proof: Controllability Test (1) ⇒ (3)

We know that controllability over the interval [0, T] means that the range space of the reachability map,

$$L_r(u)=\int_0^T e^{A(T-\tau)}\ Bu(\tau)d\tau$$

Must be the whole state space, R_n.

Notice that,

$$W_{r,T} :=\int_0^T e^{At}\ BB^* e^{A^*t}(\tau)dt=L_r\ L_r^*$$

Where,

L_r^* is the ad joint of L_r.

From the finite rank linear map theorem, $R\ (L_r)$ = R(Lr L_r^*), but $L_r\ L_r^*$ is nothing but $W_{r,T}$.

Proof of (3) ⇒ (4)

We will show this by showing "not (4) ⇒ not (3)". Suppose that (4) is not true. Then, there exists a 1 × n vector v^T so that,

$$v^T\ B=v^T\ AB=v^TA^2B...=v^TA^{n-1}B=0$$

Consider $v^T\ W_{r,T\upsilon}=\int_0^T \left\|\upsilon^T\ e^{At}\ B\right\|_2^2 dt.$

Since $e^{At}=I+At+\frac{A^2t^2}{2}+\frac{A^{n-1}\ t^{\ n-1}}{n-1!}+...$ and by the Cayley Hamilton Theorem, A^k is a linear combination of I, A,... A^{n-1}, therefore, $v^Te^{AT}\ B\ =\ 0$ for all t. Hence, $v^T\ W_{r,T}\,v\ =\ 0$ or $W_{r,T}$ is not full rank.

Proof of (3) ⇒ (2)

We will show this by showing "not (2) ⇒ not (3)".

If (2) is not true, then there exists 1 × n vector v^T so that $\tilde{o}^T\ W_{\tilde{o},T}\ =\int_0^T \left\|\tilde{o}^T\ e^{At}\ B\right\|_2^2 dt=0$

Because e^{At} is continuous in t, this implies that $v^T\ e^{AT}\ B = 0$ for all t Є [0, T].

Hence, the al time derivatives of $v^T e^{AT} B = 0$ for all $t \in [0, T]$. In particular, at $t = 0$,

$$\tilde{o}^T e^{At}B\Big|_{t=0} = \tilde{o}^T B = 0$$

$$\tilde{o}^T \frac{d}{dt} e^{At}B\Big|_{t=0} = \tilde{o}^T AB = 0$$

$$\tilde{o}^T \frac{d^2}{dt^2} e^{At}B\Big|_{t=0} = \tilde{o}^T A^2B = 0$$

$$\vdots$$

$$\tilde{o}^T \frac{d^{n-1}}{dt^{n-1}} e^{At}B\Big|_{t=0} = \tilde{o}^T A^{n-1}B = 0$$

Hence,

$$\tilde{o}^T \left(B\, AB\, A^2B \ldots A^{n-1} B\right) = \left(0\ 0 \ldots 0\right)$$

i.e., The controllability matrix of (A, B) is not full rank.

Proof of (4) ⇒ (5)

We will now show "not (5) implies not (4)". Suppose (5) is not true, so that there exists a $1 \times n$ vector v and $\lambda \in C$, $\tilde{o}^T(\lambda I - A) = 0$, $\tilde{o}^T B = 0$ for some λ.

Then,

$$\tilde{o}^T A = \lambda \tilde{o}^T, \ \tilde{o}^T A^2 = \lambda^2 \tilde{o}^T \text{ etc.}$$

Bccausc $v = 0$.

By assumption, we have $\tilde{o}^T AB = \lambda \tilde{o}^T B = 0$, $\tilde{o}^T A^2 B = \lambda^2 \tilde{o}^T B = 0$ etc.

Hence,

$$v^T B = \upsilon^T AB = \upsilon^T A^2B = \ldots \upsilon^T A^{n-1}B = 0.$$

Therefore, the controllability is not full rank.

Proof of (2) ⇒ (1)

Note that (2) ⇒ (1) is obvious.

To see that (1) ⇒ (2), notice that the controllability matrix C does not depend on T, so that the controllability does not depend on duration of time interval. Since the system is time invariant, it does not depend on the initial time (t_0).

Remarks:

- Notice that the controllability matrix is not a function of the time interval. Hence, if the LTI system is controllable over some interval, it is controllable over any (non-zero) interval with result of linear time varying system.
- Because of the above fact, we often say that the pair (A, B) is controllable.
- Controllability test can be done by examining A and B without computing the grammian. The test in (4) is attractive in that it enumerates the vectors in the controllability subspace. However, numerically, since it involves power of A, numerical stability needs to be considered.
- The PBH Test in (5) or the Haut us test for short, involves simply checking the condition at the eigenvalues. It is because for (sI – A, B) to have rank less than n, s must be an eigenvalue.
- The range space of the controllability matrix is of special interests. It is called the controllable subspace and is the set of all states that can be reached from zero-initial condition. This is A–invariant.
- Using the basis for the controllable subspace as part of the basis for Rn, the observability question deals with the question of whether one can determine the initial state x (0). Given that the input u(t), t Є [0, T] and the output y(t), t Є [0, T].

For observability, the null space of the observability map,

$L_o : x®y(\bullet) = j(\bullet,o)x, \tau Є$ [0, T] must be trivial (i.e. only contains 0). Otherwise, if $L_o(xn)(t) = y(t) = 0$, for all t Є [0, T], then for any α Є R,

$$y(t) = \phi(t,o)x = \phi(t,o)(x + \alpha \times n).$$

Hence, we cannot distinguish the various initial conditions from the output. Just like controllability, it is inconvenient to check the rank of Lo which is tall and thin.

Instead we can check the rank and the null s pace of the observability grammian given by,

$$W_o, T = L_o^* \; L_o$$

Where L_o^* is the ad joint of the observability map.

Proposition

$$\text{Null}(L_o) = \text{Null}(L_o^* \; L_o).$$

Proof:

- Let $x \in Nul(L_o)$ so, $L_o x = o$. Then, clearly, $L_o^T L_o x = L^T o$. Therefore, Null $Null(L_o) \subset Null(L_o^* L_o)$.
- Let $x \in Nul(L_o^* L_o)$. Then, $x^T L_o^* L_o x = o$. Or, $(L_o x)^T (L_o x) = o$. This can only be true if $L_o x = o$. Therefore, $Null(L_o^* L_o) \subset Null(L_o)$.

Observability Tests for Continuous Time LTI systems

Theorem

For the LTI continuous time system, the followings are equivalent:

- The system is observable over the interval [o, T].
- The observability grammian,

  $$W_{o,T} = \int_o^T e^{A^* t} C^* C e^{At} dt \text{ is full rank n.}$$

- The observability matrix $\begin{pmatrix} C \\ CA \\ CA^2 \\ \vdots \\ CA^{n-1} \end{pmatrix}$ has rank n. Notice that the observability matrix has dimension $np \times n$ where p is the dimension of the output vector y.
- For each s Є C, the matrix $\begin{pmatrix} sI - A \\ C \end{pmatrix}$ has rank n.

Proof

The proof is similar to the controllable case. Some differences are that instead of considering 1 × n vector v^T multiplying on the left hand sides of the controllability matrix and the grammians, we consider n × 1 vector multiplying on the RHS of the observability matrix and the grammian etc.

Also, instead of considering the range space of the controllability matrix, we consider the NULL space of the observability matrix. Its null space is also A-invariant. Hence if the observability matrix is not full rank, then using basis for its null space as the last k basis vectors of Rn, the system can be represented as,

$$\dot{z} = \begin{pmatrix} \tilde{A}_{11} & o \\ \tilde{A}_{21} & \tilde{A}_{22} \end{pmatrix} z + \begin{pmatrix} \tilde{B}_1 \\ \tilde{B}_2 \end{pmatrix} u$$

$$y = \begin{pmatrix} \tilde{C} & o \end{pmatrix} z$$

Where,

$C = \begin{pmatrix} \tilde{C} & o \end{pmatrix} T^{-1}$ and $A = T \begin{pmatrix} \tilde{A}_{11} & \tilde{A}_{12} \\ o & \tilde{A}_{22} \end{pmatrix} T^{-1}$ the dim of $\tilde{A}_{22}$ is non-zero.

Remarks

- Again, observability of a LTI system does not depend s on the time interval. In reality, when more data is available, one can do more averaging to eliminate the effects of noise (e.g. using the Least square i. e., Kalman Filter approach).
- The subspace of particular interest is the null space of the controllability matrix. An initial state lying in this set will generate identically zero-input response. This subspace is termed as the unobservable subspace.

Observability Tests for Discrete time LTI systems

The tests for observability of the discrete time system is given similarly by the following theorem.

Theorem

For the discrete time system, the followings are equivalent:

- The system is observable over the interval [o, T] for some $T \geq n$.
- The observability grammian,

  $$W_{o,T} = \sum_{K=o}^{T-1} A^{*k} C^* CA^k \text{ is full rank n.}$$

- The observability matrix $\begin{pmatrix} C \\ CA \\ CA^2 \\ \vdots \\ CA^{n-1} \end{pmatrix}$ has rank n. Notice that the observability matrix has dimension $np \times n$ where p is the dimension of the output vector y.
- For each z Є C, the matrix $\begin{pmatrix} sI - A \\ C \end{pmatrix}$ has rank n.

The controllability matrix can have the following interpretation:

Zero- input response is given by $\begin{pmatrix} y(0) \\ y(1) \\ y(2) \\ \vdots \\ y(n-1) \end{pmatrix} = \begin{pmatrix} C \\ CA \\ CA^2 \\ \vdots \\ CA^{n-1} \end{pmatrix} x(0).$

Thus, clearly if the observability matrix is full rank, one can reconstruct x(0) from measurements of y(0), ... , y(n - 1). One might think that by increasing the number of output measurement (e.g. y(n)), the system can become observable. However, because of Cayley-Hamilton theorem, y(n) = $CA^{n}x(0)$ can be expressed as $\sum_{i=0}^{n-1} a_i CA^i x(0)$.

Problems

1. Let us check whether the below system is completely observable or not.

$$A\begin{bmatrix} 0 & 1 & 0 \\ 0 & 0 & 1 \\ -6 & -11 & -6 \end{bmatrix}, B=\begin{bmatrix} 0 \\ 0 \\ 1 \end{bmatrix} \text{ and } C=\begin{bmatrix} 4 & 5 & 1 \end{bmatrix}$$

Solution:

Given:

$$A\begin{bmatrix} 0 & 1 & 0 \\ 0 & 0 & 1 \\ -6 & -11 & -6 \end{bmatrix}, B=\begin{bmatrix} 0 \\ 0 \\ 1 \end{bmatrix} \text{ and } C=\begin{bmatrix} 4 & 5 & 1 \end{bmatrix}$$

Formula to be used:

Condition for the system to be completely observable is,

$$\begin{vmatrix} C \\ CA \\ CA^2 \\ \bullet \\ \bullet \\ CA^{n-1} \end{vmatrix} \text{ is of rank 'n'}$$

$$CA=\begin{bmatrix} 4 & 5 & 1 \end{bmatrix}*\begin{bmatrix} 0 & 1 & 0 \\ 0 & 0 & 1 \\ -6 & -11 & -6 \end{bmatrix}=\begin{bmatrix} 5 & -7 & 5 \end{bmatrix}$$

$$CA^2 = \begin{bmatrix} 4 & 5 & 1 \end{bmatrix} * \begin{bmatrix} 0 & 1 & 0 \\ 0 & 0 & 1 \\ 6 & -11 & -6 \end{bmatrix} * \begin{bmatrix} 0 & 1 & 0 \\ 0 & 0 & 1 \\ -6 & -11 & -6 \end{bmatrix} = \begin{bmatrix} 1 & -1 & -1 \end{bmatrix}$$

$$\begin{bmatrix} C \\ CA \\ CA^2 \end{bmatrix} = \begin{bmatrix} 4 & 5 & 1 \\ 5 & -7 & -1 \\ 1 & 5 & -1 \end{bmatrix}$$

The rank of the matrix is < 3, since $\Delta = 0$.

Therefore, the system is completely not observable.

2. Let us evaluate the observability of the system using Gilbert's test if the state modal matrix is $A = \begin{bmatrix} 0 & 1 & 0 \\ 0 & 0 & 1 \\ 0 & -2 & -3 \end{bmatrix}$, $B = \begin{bmatrix} 0 \\ 0 \\ 1 \end{bmatrix}$ and C = [3 4 1].

Solution:

Given:

$$A = \begin{bmatrix} 0 & 1 & 0 \\ 0 & 0 & 1 \\ 0 & -2 & -3 \end{bmatrix}, B = \begin{bmatrix} 0 \\ 0 \\ 1 \end{bmatrix} \text{ and } C = \begin{bmatrix} 3 & 4 & 1 \end{bmatrix}$$

Formula to be used:

The composite matrix for observability,

$$Q_o = [C^T \; AT \; C^T \; (AT)^2 \; C^T]$$

Observability

$$A = \begin{bmatrix} 0 & 1 & 0 \\ 0 & 0 & 1 \\ 0 & -2 & -3 \end{bmatrix}, B = \begin{bmatrix} 0 \\ 0 \\ 1 \end{bmatrix} \text{ and } C = \begin{bmatrix} 3 & 4 & 1 \end{bmatrix}$$

$$C^T = \begin{bmatrix} 3 \\ 4 \\ 1 \end{bmatrix}$$

$$A^T = \begin{bmatrix} 0 & 0 & 0 \\ 1 & 0 & -2 \\ 0 & 1 & -3 \end{bmatrix}$$

$$A^T C^T = \begin{bmatrix} 0 & 0 & 0 \\ 1 & 0 & -2 \\ 0 & 1 & -3 \end{bmatrix} \begin{bmatrix} 3 \\ 4 \\ 1 \end{bmatrix} = \begin{bmatrix} 0 \\ 1 \\ 1 \end{bmatrix}$$

$$\left(A^T\right)^2 = \begin{bmatrix} 0 & 0 & 0 \\ 1 & 0 & -2 \\ 0 & 1 & -3 \end{bmatrix} \begin{bmatrix} 0 & 0 & 0 \\ 1 & 0 & -2 \\ 0 & 1 & -3 \end{bmatrix} = \begin{bmatrix} 0 & 0 & 0 \\ 0 & -2 & 6 \\ 1 & -3 & 7 \end{bmatrix}$$

$$\left(A^T\right)^2 C^2 = \begin{bmatrix} 0 & 0 & 0 \\ 0 & -2 & 6 \\ 1 & -3 & 7 \end{bmatrix} \begin{bmatrix} 3 \\ 4 \\ 1 \end{bmatrix} = \begin{bmatrix} 0 \\ -2 \\ -2 \end{bmatrix}$$

The composite matrix for observability is given by,

$$Q_o = [C^T \; A^T C^T \; \left(A^T\right)2 \; C^T]$$

$$= \begin{bmatrix} 3 & 0 & 0 \\ 4 & 1 & -2 \\ 1 & 1 & -2 \end{bmatrix}$$

Determinant of $Q_o = AQ_o$

$$= \begin{bmatrix} 3 & 0 & 0 \\ 4 & 1 & -2 \\ 1 & 1 & -2 \end{bmatrix}$$

$$= 3(-2+2) - 0 + 0 = 0$$

$$|Q_o| = 0.$$

Therefore, its rank ≠ 3. Hence, the system is not observable.

Permissions

All chapters in this book are published with permission under the Creative Commons Attribution Share Alike License or equivalent. Every chapter published in this book has been scrutinized by our experts. Their significance has been extensively debated. The topics covered herein carry significant information for a comprehensive understanding. They may even be implemented as practical applications or may be referred to as a beginning point for further studies.

We would like to thank the editorial team for lending their expertise to make the book truly unique. They have played a crucial role in the development of this book. Without their invaluable contributions this book wouldn't have been possible. They have made vital efforts to compile up to date information on the varied aspects of this subject to make this book a valuable addition to the collection of many professionals and students.

This book was conceptualized with the vision of imparting up-to-date and integrated information in this field. To ensure the same, a matchless editorial board was set up. Every individual on the board went through rigorous rounds of assessment to prove their worth. After which they invested a large part of their time researching and compiling the most relevant data for our readers.

The editorial board has been involved in producing this book since its inception. They have spent rigorous hours researching and exploring the diverse topics which have resulted in the successful publishing of this book. They have passed on their knowledge of decades through this book. To expedite this challenging task, the publisher supported the team at every step. A small team of assistant editors was also appointed to further simplify the editing procedure and attain best results for the readers.

Apart from the editorial board, the designing team has also invested a significant amount of their time in understanding the subject and creating the most relevant covers. They scrutinized every image to scout for the most suitable representation of the subject and create an appropriate cover for the book.

The publishing team has been an ardent support to the editorial, designing and production team. Their endless efforts to recruit the best for this project, has resulted in the accomplishment of this book. They are a veteran in the field of academics and their pool of knowledge is as vast as their experience in printing. Their expertise and guidance has proved useful at every step. Their uncompromising quality standards have made this book an exceptional effort. Their encouragement from time to time has been an inspiration for everyone.

The publisher and the editorial board hope that this book will prove to be a valuable piece of knowledge for students, practitioners and scholars across the globe.

Index

www.ingramcontent.com/pod-product-compliance
Lightning Source LLC
Chambersburg PA
CBHW061956190326
41458CB00009B/2885
9781647254308